Holt Mathematics

Course 3

Problem Solving Workbook

HOLT, RINEHART AND WINSTON

A Harcourt Education Company

Orlando • **Austin** • New York • San Diego • London

Printed in the United States of America

ISBN 0-03-079751-9

15 16 17 18 19 20 1421 15 14 13 12 11

4500312211

CONTENTS

CONTENTS, *CONTINUED*

CONTENTS, *CONTINUED*

CONTENTS, *CONTINUED*

Chapter 13

Chapter 14

Name ______________________ Date __________ Class __________

LESSON 1-1

Problem Solving

Variables and Expressions

Write the correct answer.

1. If l is the length of a room and w is the width, then lw can be used to find the area of the room. Find the area of a room with $l = 10$ ft and $w = 15$ ft.

____

2. If l is the length of a room and w is the width, $2l + 2w$ can be used to find the perimeter of the room. Find the perimeter of a room with $l = 12$ ft and $w = 16$ ft.

____

3. Jaime earns 20% commission on her sales. If s is her total sales, then $0.2s$ can be used to find the amount she earns in commission. Find her commission if her sales are $1200.

____

4. If p is the regular hourly rate of pay, then $1.5p$ can be used to find the overtime rate of pay. Find the overtime rate of pay if the regular hourly rate of pay is $6.00 per hour.

____

Choose the letter for the best answer.

5. A plumber charges a fee of $75 per service call plus $15 per hour. If h is the number of hours the plumber works, then $75 + 15h$ can be used to find the total charges. Find the total charges if the plumber works 2.5 hours.

A $37.50
B $112.50
C $225
D $1127.50

6. Tickets to the movies cost $4 for students and $6 for adults. If s is the number of students and a is the number of adults, $4s + 6a$ can be used to find the cost of the tickets. Find the cost of the tickets for 3 students and 2 adults.

F $15
G $17
H $24
J $26

7. If c is the number of cricket chirps in a minute, then the expression $0.25c + 20$ can be used to estimate the temperature in degrees Farenheit. If there are 92 cricket chirps in a minute, find the temperature.

A 43 degrees
B 33 degrees
C 102 degrees
D 75 degrees

8. Flowers are sold in flats of 6 plants each. If f is the number of flats, then $6f$ can be used to find the number of flowers. Find the number of flowers in 18 flats.

F 3 flowers
G 108 flowers
H 24 flowers
J 12 flowers

Name ______________________ Date ___________ Class ___________

LESSON **1-2**

Problem Solving

Algebraic Expressions

Write the correct answer.

1. Morton bought 15 new books to add to his collection of books b. Write an algebraic expression to evaluate the total number of books in Morton's collection if he had 20 books in his collection.

____

2. Paul exercises m minutes per day 5 days a week. Write an algebraic expression to evaluate how many minutes Paul exercises each week if he exercises 45 minutes per day.

____

3. Helen bought 3 shirts that each cost s dollars. Write an algebraic expression to evaluate how much Helen spent in all if each shirt cost $22.

____

4. Claire makes b bracelets to divide evenly among four friends and herself. Write an algebraic expression to evaluate the number of bracelets each person will receive if Claire makes 15 bracelets.

____

Choose the letter for the best answer.

5. Jonas collects baseball cards. He has 245 cards in his collection. For his birthday, he received r more cards, then he gave his brother g cards. Which algebraic expression represents the total number of cards he now has in his collection?

   A $245 + r + g$

   B $245 - r - g$

   C $245 + r - g$

   D $r + g - 245$

6. Monique is saving money for a computer. She has m dollars saved. For her birthday, her dad doubled her money, but then she spent s dollars on a shirt. Which algebraic expression represents the amount of money she has now saved for her computer?

   F $m + 2 - s$

   G $2m - s$

   H $2m + s$

   J $m + 2s$

7. Which algebraic expression represents the number of years in m months?

   A $12m$

   B $\frac{m}{12}$

   C $12 + m$

   D $12 - m$

8. Which algebraic expression represents how many minutes are in h hours?

   F $60h$

   G $\frac{h}{60}$

   H $h + 60$

   J $h - 60$

Name ____________________ Date __________ Class __________

LESSON 1-3

Problem Solving

Integers and Absolute Value

Write the correct answer.

1. In Africa, Lake Asal reaches a depth of −153 meters. In Asia, the Dead Sea reaches a depth of −408 meters. Which reaches a greater depth, Lake Asal or the Dead Sea?

____

2. Jeremy's scores for four golf games are: −1, 2, −3, and 1. Order his golf scores from least to greatest.

____

3. The lowest point in North America is Death Valley with an elevation of −282 feet. South America's lowest point is the Valdes Peninsula with an elevation of −131 feet. Which continent has the lowest point?

____

4. Two undersea cameras are taking time lapse photos in a coral reef. The first camera is mounted at −45 feet. The second camera is mounted at −25. Which camera is closer to the surface?

____

Use the table to answer Exercises 5–7. Choose the letter of the best answer.

State Low Temperature Records

State	Temperature (°F)
Alabama	−27
Indiana	−36
Massachusetts	−35
Texas	−23

5. Which state had the coldest temperature?

   A Alabama　　**C** Massachusetts
   B Indiana　　**D** Texas

6. Which is the greatest temperature listed?

   F −27°F　　**H** −35°F
   G −36°F　　**J** −23°F

7. The lowest temperature recorded in Connecticut was between the lowest temperatures recorded in Alabama and Massachusetts. Which could be the lowest temperature recorded in Connecticut?

   A −40°F　　**C** −32°F
   B −37°F　　**D** −40°F

Name ______________________ Date __________ Class __________

LESSON **1-4**

Problem Solving

Adding Integers

Use the following information for Exercises 1–3. In golf, par 73 means that a golfer should take 73 strokes to finish 18 holes. A score of 68 is 5 under par, or −5. A score of 77 is 4 over par, or +4.

1. Use integers to write Tiger Woods's score for each round as over or under par.

____

2. Add the integers to find Tiger Woods's overall score.

____

3. Was Tiger Woods's overall score over or under par?

____

Tiger Woods's Scores
Mercedes Championship
January 6, 2002
Par 73 course

Round	Score
1	68
2	74
3	74
4	65

Choose the letter for the best answer.

4. At 9:00 A.M., the temperature was −15°. An hour later, the temperature had risen 7°. What is the temperature now?

   A −22° **C** −8°
   B 8° **D** 22°

5. Sandra is reviewing her savings account statement. She withdrew amounts of $35, $20, and $15. She deposited $65. If her starting balance was $657, find the new balance.

   F $652 **H** $662
   G $522 **J** $507

6. During a possession in a football game, the Vikings gained 22 yards, lost 15 yards, gained 3 yards, gained 20 yards and lost 5 yards. At the end of the possession, how many yards had they lost or gained?

   A gained 43 yards
   B lost 43 yards
   C lost 25 yards
   D gained 25 yards

7. A submarine is cruising at 40 m below sea level. The submarine ascends 18 m. What is the submarine's new location?

   F 58 m below sea level
   G 22 m below sea level
   H 18 m below sea level
   J 12 m below sea level

Name ______________________ Date __________ Class __________

LESSON 1-5

Problem Solving

Subtracting Integers

Write the correct answer.

1. In Fairbanks, Alaska, the average January temperature is −13°F, while the average April temperature is 30°F. What is the difference between the average temperatures?

   ____

2. The highest point in North America is Mt. McKinley, Alaska, at 20,320 ft above sea level. The lowest point is Death Valley, California, at 282 ft below sea level. What is the difference in elevations?

   ____

3. The temperature fell from 44°F to −56°F in 24 hours in Browning, Montana, on January 23–24, 1916. By how many degrees did the temperature change?

   ____

4. The boiling point of chlorine is −102°C, while the melting point is −34°C. What is the difference between the melting and boiling points of chlorine?

   ____

Use the table below to answer Exercises 5–7. The table shows the first and fifth place finishers in a golf tournament. In golf, the winner has the lowest total for all five rounds. Choose the letter for the best answer.

Bob Hope Chrysler Classic
January 20, 2002

Round	J. Kelly	P. Mickelson
1	−8	−8
2	−3	−5
3	−7	−2
4	−4	−7
5	−5	−8

5. By how many points did Mickelson beat Kelly in Round 2?

   A 2 **C** 5
   B 3 **D** 8

6. By how many points did Kelly beat Mickelson in Round 3?

   F 2 **H** 5
   G 3 **J** 9

7. Who won the Bob Hope Chrysler Classic and how many points difference was there between first and fifth place?

   A Kelly; 4 **C** Kelly; 3
   B Mickelson; 4 **D** Mickelson; 3

Name ______________________________ Date ____________ Class ____________

LESSON 1-6

Problem Solving

Multiplying and Dividing Integers

Write the correct answer.

1. A submersible started at the surface of the water and was moving down at −12 meters per minute toward the ocean floor. The submersible traveled at this rate for 32 minutes before coming to rest on the ocean floor. What is the depth of the ocean floor?

____

2. For the first week in January, the daily high temperatures in Bismarck, North Dakota, were 7°F, −10°F, −10°F, −7°F, 8°F, 12°F, and 14°F. What was the average daily high temperature for the week?

____

3. Sally went golfing and recorded her scores as −2 on the first hole, −2 on the second hole, and 1 on the third hole. What is her average for the first three holes?

____

4. The ocean floor is at −96 m. Tom has reached −15 m. If he continues to move down at −3 m per minute, how far will he be from the ocean floor after 7 minutes?

____

Use the table below to answer Exercises 5–7. Choose the letter for the best answer.

Calories Consumed or Burned

Food or Exercise	Calories
Apple	125
Pepperoni pizza (slice)	181
Hamburger	425
Basketball (1hr)	−545
In-line skating (1 hr)	−477
Frisbee (1 hr)	−205

5. What is the caloric impact of 2 hours of in-line skating?

A −477 Cal **C** −583 Cal
B −479 Cal **D** −954 Cal

6. What is the caloric impact of eating a hamburger and then playing Frisbee for 3 hours?

F 220 Cal **H** 190 Cal
G −190 Cal **J** −220 Cal

7. Tim plays basketball for 1 hour, skates for 5 hours, and plays Frisbee for 4 hours. What is the average amount of calories Tim burns per hour?

A −375 Cal **C** −545 Cal
B −1250 Cal **D** −409 Cal

Name ______________________ Date __________ Class __________

LESSON 1-7

Problem Solving

Solving Equations by Adding or Subtracting

Write the correct answer.

1. The 1954 elevation of Mt. Everest was 29,028 ft. In 1999, that elevation was revised to be 29,035 ft. Write an equation to find the change c in elevation of Mt. Everest.

____

2. The difference between the boiling and melting points of fluorine is 32°C. If the boiling point of fluorine is −188°C, write an equation and solve to find the melting point m of fluorine.

____

____

3. Lisa sold her old bike for $140 less than she paid for it. She sold the bike for $85. Write and solve an equation to find how much Lisa paid for her bike.

____

____

4. The average January temperature in Fairbanks, Alaska, is −13°F. The April average is 43°F higher than the January average. Write an equation to find the average April temperature.

____

____

Choose the letter for the best answer.

5. A survey found that female teens watched 3 hours of TV per week less than male teens. The female teens reported watching an average of 18 hours of TV. Find the number of hours h the male teens watched.

A $h = 6$ **C** $h = 18$

B $h = 15$ **D** $h = 21$

6. It costs about $125 more per year to feed a hamster than it does to feed a bird. If it costs $256 per year to feed a hamster, find the cost c to feed a bird.

F $c = \$131$ **H** $c = \$256$

G $c = \$125$ **J** $c = \$381$

7. Naples, Florida, is the second fastest growing U.S. metropolitan area. From 1990 to 2000, the population increased by 99,278. If the 2000 population was 251,377, find the population p in 1990.

A $p = 253{,}377$ **C** $p = 249{,}377$

B $p = 350{,}655$ **D** $p = 152{,}099$

8. In 1940, the life expectancy for a female was 65 years. In 1999, the life expectancy for a female was 79 years. Find the increase in the life expectancy for females.

F 14 yrs **H** −14 yrs

G 1.2 yrs **J** 144 yrs

Name ______________________ Date __________ Class __________

LESSON 1-8

Problem Solving

Solving Equations by Multiplying or Dividing

Write the correct answer.

1. Brett is preparing to participate in a 250-kilometer bike race. He rides a course near his house that is 2 km long. Write an equation to determine how many laps he must ride to equal the distance of the race.

____

2. The average life span of a duck is 10 years, which is one year longer than three times the average life span of a guinea pig. Write and solve an equation to determine the lifespan of a guinea pig.

____

3. The speed of a house mouse is one-fourth that of a giraffe. If a house mouse can travel at 8 mi/h, what is the speed of a giraffe? Write an equation and solve.

____

____

4. In 2005, the movie with the highest box office sales was *Titanic,* which made about 3 times the box office sales of *Charlie and the Chocolate Factory*. If *Titanic* made about $600 million, about how much did *Charlie and the Chocolate Factory* make? Write an equation and solve.

____

____

Choose the letter for the best answer.

5. Farmland is often measured in acres. A farm that is 1920 acres covers 3 square miles. Find the number of acres *a* in one square mile.

A 9 acres
B 213 acres
C 640 acres
D 4860 acres

6. When Maria doubles a recipe, she uses 8 cups of flour. How many cups of flour are in the original recipe?

F 2 cups
G 4 cups
H 8 cups
J 16 cups

7. The depth of water is often measured in fathoms. A fathom is six feet. If the maximum depth of the Gulf of Mexico is 2395 fathoms, what is the maximum depth in feet?

A 14,370 ft
B 98,867 ft
C 29,250 ft
D 175,464 ft

8. Four times as many pet birds have lived in the White House as pet goats. Sixteen pet birds have lived in the White House. How many pet goats have there been?

F 4
G 20
H 12
J 64

Name ______________________ Date __________ Class __________

LESSON 1-9

Problem Solving

Introduction to Inequalities

Use the table.

1. Write an inequality that compares the population p of Los Angeles to the population of New York.

____

2. Write an inequality that compares the population p of Los Angeles to the population of Chicago.

____

Top 3 U.S. Cities by Population 2000

Rank	City	Population
1	New York	8,008,278
2	Los Angeles	p
3	Chicago	2,896,016

Write the correct answer.

3. Paul wants to ride his bike at least 30 miles this week to train for a race. He has already ridden 18 miles. How many more miles should Paul ride this week?

____

4. To avoid a service charge, Jose must keep more than $500 in his account. His current balance is $536, but he plans to write a check for $157. Find the amount of the deposit d Jose must make to avoid a service charge.

____

Choose the letter for the best answer.

5. Mia wants to spend no more than $10 on an ad in the paper. The first 10 words cost $3. Find the amount of money m she has left to spend on the ad.

   A $m \geq 7$ **C** $m \leq 7$
   B $m \leq 13$ **D** $m \geq 13$

6. An auto shop estimates parts and labor for a repair will cost less than $200. Parts will cost $59. Find the maximum cost c of the labor.

   F $c < \$141$ **H** $c > \$141$
   G $c < \$259$ **J** $c > \$259$

7. To advance to the next level of a competition, Rachel must earn at least 180 points. She has already earned 145 points. Find the number of points p she needs to advance to the next level of the competition.

   A $p \leq 35$ **C** $p \geq 35$
   B $p \leq 325$ **D** $p \geq 325$

8. The Conway's hiked more than 25 miles on their backpacking trip. If they hiked 8 miles on their last day, find how many miles m they hiked on the rest of the trip.

   F $m > 17$ **H** $m < 17$
   G $m > 33$ **J** $m < 33$

Name ______________________ Date ____________ Class ____________

LESSON 2-1

Problem Solving

Rational Numbers

Write the correct answer.

1. Fill in the table below which shows the sizes of drill bits in a set.

2. Do the drill bit sizes convert to repeating or terminating decimals?

____

13-Piece Drill Bit Set

Fraction	Decimal	Fraction	Decimal	Fraction	Decimal
$\frac{1}{4}$"		$\frac{11}{64}$"		$\frac{3}{32}$"	
$\frac{15}{64}$"		$\frac{5}{32}$"		$\frac{5}{64}$"	
$\frac{7}{32}$"		$\frac{9}{64}$"		$\frac{1}{16}$"	
$\frac{13}{64}$"		$\frac{1}{8}$"			
$\frac{3}{16}$"		$\frac{7}{64}$"			

Use the table at the right that lists the world's smallest nations. Choose the letter for the best answer.

World's Smallest Nations

Nation	**Area** (square miles)
Vatican City	0.17
Monaco	0.75
Nauru	8.2

3. What is the area of Vatican City expressed as a fraction in simplest form?

A $\frac{8}{50}$

B $\frac{4}{25}$

C $\frac{17}{1000}$

D $\frac{17}{100}$

4. What is the area of Monaco expressed as a fraction in simplest form?

F $\frac{75}{100}$

G $\frac{15}{20}$

H $\frac{3}{4}$

J $\frac{2}{3}$

5. What is the area of Nauru expressed as a mixed number?

A $8\frac{1}{50}$

B $8\frac{2}{50}$

C $8\frac{2}{100}$

D $8\frac{1}{5}$

6. The average annual precipitation in Miami, FL is 57.55 inches. Express 57.55 as a mixed number.

F $57\frac{11}{20}$

G $57\frac{55}{1000}$

H $57\frac{5}{100}$

J $57\frac{1}{20}$

7. The average annual precipitation in Norfolk, VA is 45.22 inches. Express 45.22 as a mixed number.

A $45\frac{11}{50}$

B $45\frac{22}{1000}$

C $45\frac{11}{20}$

D $45\frac{1}{5}$

Name ______________________ Date __________ Class __________

LESSON 2-2

Problem Solving

Comparing and Ordering Rational Numbers

Write the correct answer.

1. Carl Lewis won the gold medal in the long jump in four consecutive Summer Olympic games. He jumped 8.54 meters in 1984, 8.72 meters in 1988, 8.67 meters in 1992, 8.5 meters in 1996. Order the length of his winning jumps from least to greatest.

____

2. Scientists aboard a submarine are gathering data at an elevation of $-42\frac{1}{2}$ feet. Scientists aboard a submersible are taking photographs at an elevation of $-45\frac{1}{3}$ feet. Which scientists are closer to the surface of the ocean?

____

3. The depth of a lake is measured at three different points. Point A is −15.8 meters, Point B is −17.3 meters, and Point C is −16.9 meters. Which point has the greatest depth?

____

4. At a swimming meet, Gail's time in her first heat was $42\frac{3}{8}$ seconds. Her time in the second heat was 42.25 seconds. Which heat did she swim faster?

____

The table shows the top times in a 5 K race. Choose the letter of the best answer.

Name	Time (minutes)
Marshall	18.09
Renzo	17.38
Dan	17.9
Aaron	18.61

5. Who had the fastest time in the race?

A Marshall

B Renzo

C Dan

D Aaron

6. Which is the slowest time in the table?

F 18.09 minutes

G 17.38

H 17.9 minutes

J 18.61 minutes [*]

7. Aaron's time in a previous race was less than his time in this race but greater than Marshall's time in this race. How fast could Aaron have run in the previous race?

A 19.24 min

B 18.7 min

C 18.35 min

D 18.05 mi

Name ______________________ Date __________ Class __________

LESSON 2-3

Problem Solving

Adding and Subtracting Rational Numbers

Write the correct answer.

1. In 2004, Yuliya Nesterenko of Belarus won the Olympic Gold in the 100-m dash with a time of 10.93 seconds. In 2000, American Marion Jones won the 100-m dash with a time of 10.75 seconds. How many seconds faster did Marion Jones run the 100-m dash?

____

2. The snowfall in Rochester, NY in the winter of 1999–2000 was 91.5 inches. Normal snowfall is about 76 inches per winter. How much more snow fell in the winter of 1999–2000 than is normal?

____

3. In a survey, $\frac{76}{100}$ people indicated that they check their e-mail daily, while $\frac{23}{100}$ check their e-mail weekly, and $\frac{1}{100}$ check their e-mail less than once a week. What fraction of people check their e-mail at least once a week?

____

4. To make a small amount of play dough, you can mix the following ingredients: 1 cup of flour, $\frac{1}{2}$ cup of salt and $\frac{1}{2}$ cup of water. What is the total amount of ingredients added to make the play dough?

____

Choose the letter for the best answer.

Baseball Ticket Prices

Location	Average Price
Minnesota	$14.42
League Average	$19.82
Boston	$40.77

5. How much more expensive is it to buy a ticket in Boston than in Minnesota?

A $20.95 **C** $5.40
B $55.19 **D** $26.35

6. How much more expensive is it to buy a ticket in Boston than the league average?

F $60.59
G $20.95
H $5.40
J $26.35

7. What is the total cost of a ticket in Boston and a ticket in Minnesota?

A $55.19
B $34.24
C $60.59
D $54.19

Name ______________________ Date __________ Class __________

LESSON 2-4

Problem Solving

Multiplying Rational Numbers

Use the table at the right.

Average World Births and Deaths per Second in 2001

Births	$4\frac{1}{5}$
Deaths	1.7

1. What was the average number of births per minute in 2001?

____

2. What was the average number of deaths per hour in 2001?

____

3. What was the average number of births per day in 2001?

____

4. What was the average number of births in $\frac{1}{2}$ of a second in 2001?

____

5. What was the average number of births in $\frac{1}{4}$ of a second in 2001?

____

Use the table below. During exercise, the target heart rate is 0.5-0.75 of the maximum heart rate. Choose the letter for the best answer.

Age	Maximum Heart Rate
13	207
14	206
15	205
20	200
25	195

Source: American Heart Association

6. What is the target heart rate range for a 14 year old?

A 7–10.5

B 103–154.5

C 145–166

D 206–255

7. What is the target heart rate range for a 20 year old?

F 100–150

G 125–175

H 150–200

J 200–250

8. What is the target heart rate range for a 25 year old?

A 25–75

B 85–125

C 97.5–146.25

D 195–250

Name ______________________ Date ____________ Class ____________

LESSON 2-5

Problem Solving

Dividing Rational Numbers

Use the table at the right that shows the maximum speed over a quarter mile of different animals. Find the time is takes each animal to travel one-quarter mile at top speed. Round to the nearest thousandth.

Maximum Speeds of Animals

Animal	**Speed** (mph)
Quarter Horse	47.50
Greyhound	39.35
Human	27.89
Giant Tortoise	0.17
Three-toed sloth	0.15

1. Quarter horse

____

2. Greyhound

____

3. Human

____

4. Giant tortoise

____

5. Three-toed sloth

____

Choose the letter for the best answer.

6. A piece of ribbon is $1\frac{7}{8}$ inches long. If the ribbon is going to be divided into 15 pieces, how long should each piece be?

A $\frac{1}{8}$ in.

B $\frac{1}{15}$ in.

C $\frac{2}{3}$ in.

D $28\frac{1}{8}$ in.

7. The recorded rainfall for each day of a week was 0 in., $\frac{1}{4}$ in., $\frac{3}{4}$ in., 1 in., 0 in., $1\frac{1}{4}$ in., $1\frac{1}{4}$ in. What was the average rainfall per day?

F $\frac{9}{10}$ in.

G $\frac{9}{14}$ in.

H $\frac{7}{8}$ in.

J $4\frac{1}{2}$ in.

8. A drill bit that is $\frac{7}{32}$ in. means that the hole the bit makes has a diameter of $\frac{7}{32}$ in. Since the radius is half of the diameter, what is the radius of a hole drilled by a $\frac{7}{32}$ in. bit?

A $\frac{14}{32}$ in.

B $\frac{7}{32}$ in.

C $\frac{9}{16}$ in.

D $\frac{7}{64}$ in.

9. A serving of a certain kind of cereal is $\frac{2}{3}$ cup. There are 12 cups of cereal in the box. How many servings of cereal are in the box?

F 18

G 15

H 8

J 6

Name ______________________ Date __________ Class __________

LESSON 2-6

Problem Solving

Adding and Subtracting with Unlike Denominators

Write the correct answer.

1. Nick Hysong of the United States won the Olympic gold medal in the pole vault in 2000 with a jump of 19 ft $4\frac{1}{4}$ inches, or $232\frac{1}{4}$ inches. In 1900, Irving Baxter of the United States won the pole vault with a jump of 10 ft $9\frac{7}{8}$ inches, or $129\frac{7}{8}$ inches. How much higher did Hysong vault than Baxter?

____

2. In the 2000 Summer Olympics, Ivan Pedroso of Cuba won the Long jump with a jump of 28 ft $\frac{3}{4}$ inches, or $336\frac{3}{4}$ inches. Alvin Kraenzlein of the Unites States won the long jump in 1900 with a jump of 23 ft $6\frac{7}{8}$ inches, or $282\frac{7}{8}$ inches. How much farther did Pedroso jump than Kraenzlein?

____

3. A recipe calls for $\frac{1}{8}$ cup of sugar and $\frac{3}{4}$ cup of brown sugar. How much total sugar is added to the recipe?

____

4. The average snowfall in Norfolk, VA for January is $2\frac{3}{5}$ inches, February $2\frac{9}{10}$ inches, March 1 inch, and December $\frac{9}{10}$ inches. If these are the only months it typically snows, what is the average snowfall per year?

____

Use the table at the right that shows the average snowfall per month in Vail, Colorado.

5. What is the average annual snowfall in Vail, Colorado?

A $15\frac{13}{20}$ in.

B 153 in.

C $187\frac{1}{10}$ in.

D $187\frac{4}{5}$ in.

6. The peak of the skiing season is from December through March. What is the average snowfall for this period?

F $30\frac{19}{20}$ in.

G $123\frac{3}{5}$ in.

H $123\frac{4}{5}$ in.

J 127 in.

Average Snowfall in Vail, CO

Month	Snowfall (in.)	Month	Snowfall (in.)
Jan	$36\frac{7}{10}$	July	0
Feb	$35\frac{7}{10}$	August	0
March	$25\frac{2}{5}$	Sept	1
April	$21\frac{1}{5}$	Oct	$7\frac{4}{5}$
May	4	Nov	$29\frac{7}{10}$
June	$\frac{3}{10}$	Dec	26

Name ______________________ Date __________ Class __________

LESSON 2-7

Problem Solving

Solving Equations with Rational Numbers

Write the correct answer.

1. In the last 150 years, the average height of people in industrialized nations has increased by $\frac{1}{3}$ foot. Today, American men have an average height of $5\frac{7}{12}$ feet. What was the average height of American men 150 years ago?

____

2. Jaime has a length of ribbon that is $23\frac{1}{2}$ in. long. If she plans to cut the ribbon into pieces that are $\frac{3}{4}$ in. long, into how many pieces can she cut the ribbon? (She cannot use partial pieces.)

____

3. Todd's restaurant bill for dinner was $15.55. After he left a tip, he spent a total of $18.00 on dinner. How much money did Todd leave for a tip?

____

4. The difference between the boiling point and melting point of Hydrogen is 6.47°C. The melting point of Hydrogen is −259.34°C. What is the boiling point of Hydrogen?

____

Choose the letter for the best answer.

5. Justin Gatlin won the Olympic gold in the 100-m dash in 2004 with a time of 9.85 seconds. His time was 0.95 seconds faster than Francis Jarvis who won the 100-m dash in 1900. What was Jarvis' time in 1900?

A 8.95 seconds
B 10.65 seconds
C 10.80 seconds
D 11.20 seconds

6. The balance in Susan's checking account was $245.35. After the bank deposited interest into the account, her balance went to $248.02. How much interest did the bank pay Susan?

F $1.01
G $2.67
H $3.95
J $493.37

7. After a morning shower, there was $\frac{17}{100}$ in. of rain in the rain gauge. It rained again an hour later and the rain gauge showed $\frac{1}{4}$ in. of rain. How much did it rain the second time?

A $\frac{2}{25}$ in. **C** $\frac{21}{50}$ in.
B $\frac{1}{6}$ in. **D** $\frac{3}{8}$ in.

8. Two-third of John's savings account is being saved for his college education. If $2500 of his savings is for his college education, how much money in total is in his savings account?

F $1666.67 **H** $4250.83
G $3750 **J** $5000

Name ______________________ Date __________ Class __________

LESSON 2-8

Problem Solving

Solving Two-Step Equations

The chart below describes three different long distance calling plans. Jamie has budgeted $20 per month for long distance calls. Write the correct answer.

Plan	Monthly Access Fee	Charge per minute
A	$3.95	$0.08
B	$8.95	$0.06
C	$0	$0.10

1. How many minutes will Jamie be able to use per month with plan A? Round to the nearest minute.

____

2. How many minutes will Jamie be able to use per month with plan B? Round to the nearest minute.

____

3. How many minutes will Jamie be able to use per month with plan C? Round to the nearest minute.

____

4. Which plan is the best deal for Jamie's budget?

____

5. Nolan has budgeted $50 per month for long distance. Which plan is the best deal for Nolan's budget?

____

The table describes four different car loans that Susana can get to finance her new car. The total column gives the amount she will end up paying for the car including the down payment and the payments with interest. Choose the letter for the best answer.

Loan	Down Payment	Number of Months	Total
A	$2000	60	$19,821.20
B	$1000	48	$19,390.72
C	$0	60	$20,197.20

6. How much will Susana pay each month with loan A?

A $252.04 **C** $330.35

B $297.02 **D** $353.68

7. How much will Susana pay each month with loan B?

F $300.85 **H** $323.17

G $306.50 **J** $383.14

8. How much will Susana pay each month with loan C?

A $336.62 **C** $369.95

B $352.28 **D** $420.78

9. Which loan will give Susana the smallest monthly payment?

F Loan A **H** Loan C

G Loan B **J** They are equal

Name ________________ Date __________ Class __________

LESSON 3-1

Problem Solving

Ordered Pairs

Use the table at the right for Exercises 1–2.

1. Write the ordered pair that shows the average miles per gallon in 1990.

____

2. The data can be approximated by the equation $m = 0.30887x - 595$ where m is the average miles per gallon and x is the year. Use the equation to find an ordered pair (x, m) that shows the estimated miles per gallon in the year 2020.

____

Average Miles per Gallon

Year	Miles per Gallon
1970	13.5
1980	15.9
1990	20.2
1995	21.1
1996	21.2
1997	21.5

For Exercises 3–4 use the equation $F = 1.8C + 32$, which relates Fahrenheit temperatures F to Celsius temperatures C.

3. Write ordered pair (C, F) that shows the Celsius equivalent of 86°F.

____

4. Write ordered pair (C, F) that shows the Fahrenheit equivalent of 22°C.

____

Choose the letter for the best answer.

5. A taxi charges a $2.50 flat fee plus $0.30 per mile. Use an equation for taxi fare t in terms of miles m. Which ordered pair (m, t) shows the taxi fare for a 23-mile cab ride?

   A (23, 6.90) **C** (23, 9.40)
   B (23, 18.50) **D** (23, 64.40)

6. The perimeter p of a square is four times the length of a side s, or $p = 4s$. Which ordered pair (s, p) shows the perimeter for a square that has sides that are 5 in.?

   F (5, 1.25) **H** (5, 9)
   G (5, 20) **J** (5, 25)

7. Maria pays a monthly fee of $3.95 plus $0.10 per minute for long distance calls. Use an equation for the phone bill p in terms of the number of minutes m. Which ordered pair (m, p) shows the phone bill for 120 minutes?

   A (120, 15.95) **C** (120, 28.30)
   B (120, 474.10) **D** (120, 486.00)

8. Tickets to a baseball game cost $12 each, plus $2 each for transportation. Use an equation for the cost c of going to the game in terms of the number of people p. Which ordered pair (p, c) shows the cost for 6 people?

   F (6, 74) **H** (6, 84)
   G (6, 96) **J** (6, 102)

Name ______________________ Date ___________ Class ___________

LESSON 3-2

Problem Solving

Graphing on the Coordinate Plane

Complete the table of ordered pairs. Graph each ordered pair. Draw a line through the points. Answer the question.

1. John earns $150 per week plus 5% of his computer software sales. John's weekly pay y in terms of his sales x is $y = 150 + 0.05x$. Complete the table. How much does John get paid for $200 in sales?

____

x	y	(x, y)
0		
10		
25		
50		
100		

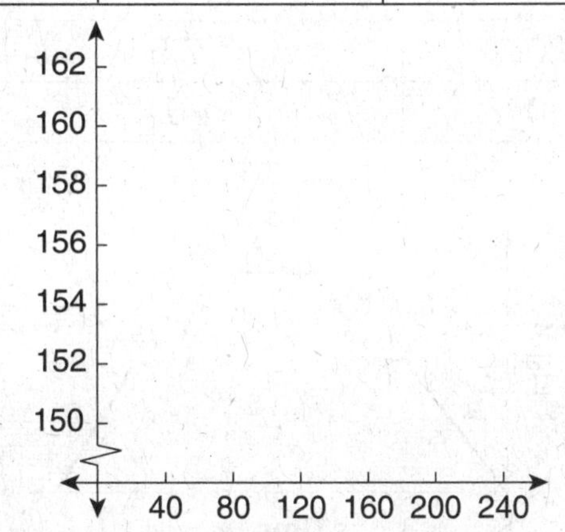

2. Margarite starts out with $100. Each week, she spends $6 to go to the movies. The amount of money y Margarite has left each week x, is $y = 100 - 6x$. How much money does she have left after 11 weeks?

____

x	y	(x, y)
0		
1		
2		
3		
4		

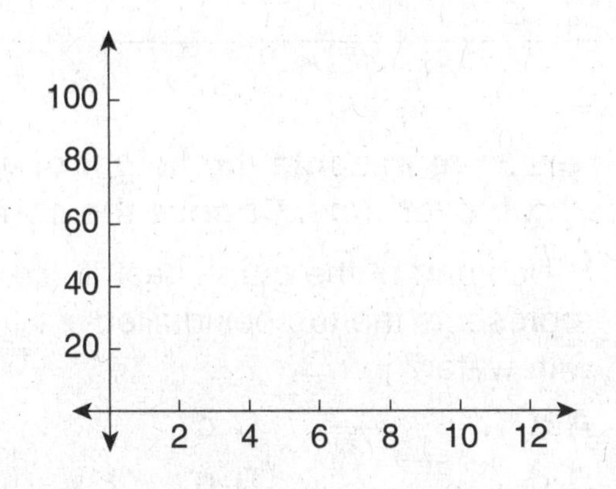

The graph at the right represents the miles traveled y in x hours. Use the graph to choose the best letter.

3. Which of the ordered pairs below represents a solution?

A (1, 65) **C** (5, 120)

B (2, 70) **D** (7, 300)

4. The graph represents a car traveling how fast?

F 60 mi/h **H** 70 mi/h

G 65 mi/h **J** 75 mi/h

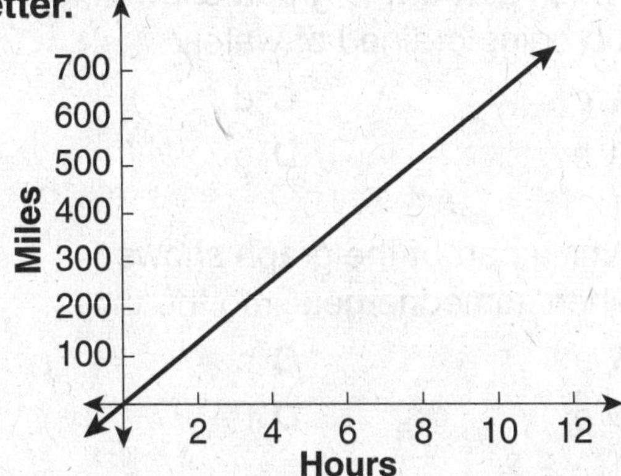

Name ______________________ Date __________ Class __________

LESSON 3-3

Problem Solving

Interpreting Graphs and Table

Tell which table corresponds to each situation.

1. Ryan walks for several blocks, and then he begins to run. After running for 10 minutes, he walks for several blocks and then stops.

____

2. Susanna starts running. After 10 minutes, she sees a friend and stops to talk. When she leaves her friend, she runs home and stops.

____

3. Mark stands on the porch and talks to a friend. Then he starts walking home. Part way home he decides to run the rest of the way, and he doesn't stop until he gets home.

____

Table 1

Time	**Speed** (mi/h)
8:00	0
8:10	3
8:20	7.5
8:30	0

Table 2

Time	**Speed** (mi/h)
8:00	3
8:10	7.5
8:20	3
8:30	0

Table 3

Time	**Speed** (mi/h)
8:00	7.5
8:10	0
8:20	7.5
8:30	0

The graph represents the height of water in a bathtub over time. Choose the correct letter.

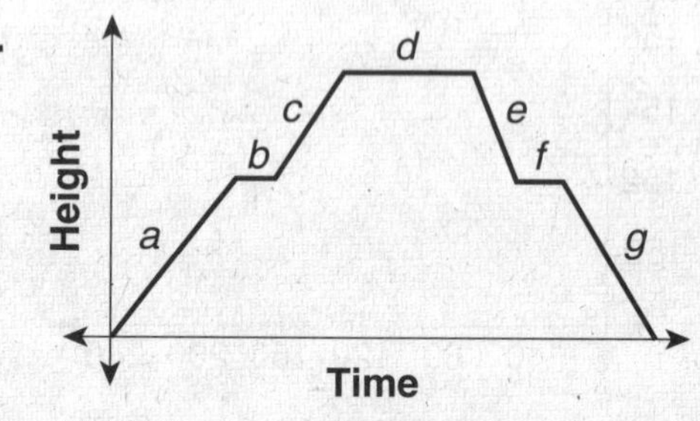

4. Which part of the graph best represents the tub being filled with water?

A a **C** c
B d **D** g

5. Which part of the graph shows the tub being drained of water?

A c **C** d
B e **D** g

6. Which part of the graph shows someone soaking in the tub?

F b **H** d
G e **J** f

7. Which part of the graph shows when someone gets into the tub?

A a **C** c
B e **D** f

8. Which parts of the graph show when the water level is not changing in the tub?

F a, b, c **H** b, d, g
G b, d, f **J** c, e, f

Name ______________________ Date __________ Class __________

LESSON 3-4

Problem Solving

Functions

A cyclist rides at an average speed of 20 miles per hour. The equation $y = 20x$ shows the distance, y, the cyclist travels in x hours.

1. Make a table for the equation and graph the equation at the right.

x	$20x$	y
0		
1		
2		
3		

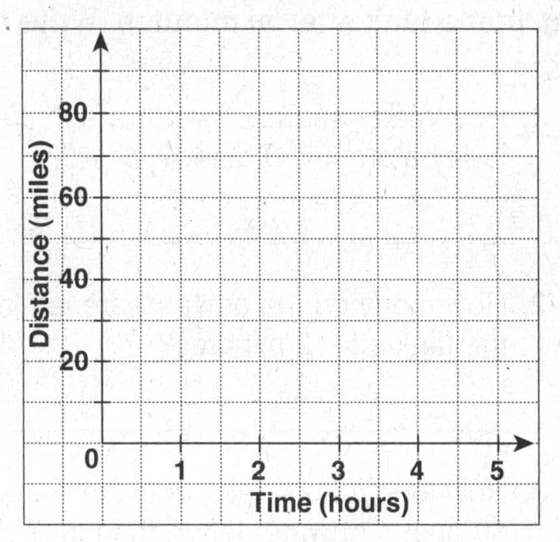

2. Is the relationship between the time and the distance the cyclist rides a function?

____

3. If the cyclist continues to ride at the same rate, about how far will the cyclist ride in 4 hours?

____

4. About how far does the cyclist ride in 1.5 hours?

____

5. If the cyclist has ridden 50 miles, about how long has the cyclist been riding?

____

The cost of renting a jet-ski at a lake is represented by the equation $f(x) = 25x + 100$ where x is the number of hours and $f(x)$ is the cost including an hourly rate and a deposit. Choose the letter for the best answer.

6. What is the domain of the function?

A $x < 0$
B $x > 0$
C $x > 25$
D $x < 100$

7. What is the range of the function?

F $f(x) > 0$
G $f(x) < 0$
H $f(x) < 25$
J $f(x) > 100$

8. How much does it cost to rent the jet-ski for 5 hours?

A \$125
B \$225
C \$385
D \$525

9. If the cost to rent the jet-ski is \$300, for how many hours is the jet-ski rented?

F 6 hours
G 8 hours
H 12 hours
J 16 hours

Name ______________________ Date ____________ Class ____________

LESSON **3-5**

Problem Solving

Equations, Tables, and Graphs

Use the graph to answer Exercises 1–4. An aquarium tank is being drained. The graph shows the number of quarts of water, *q*, in the tank after *m* minutes. Write the correct answer.

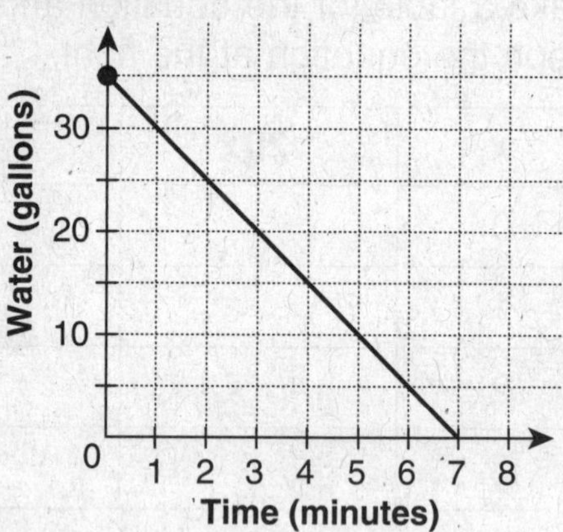

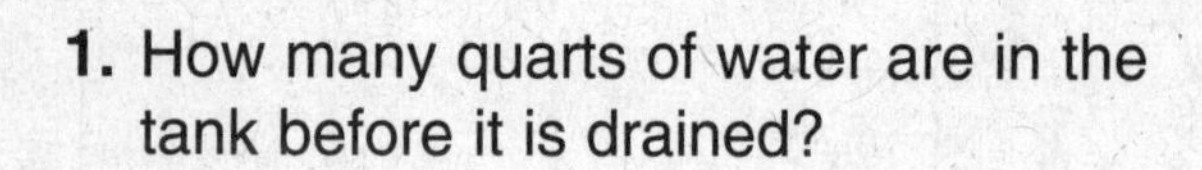

1. How many quarts of water are in the tank before it is drained?

____

2. How many quarts of water are left in the tank after 2 minutes?

____

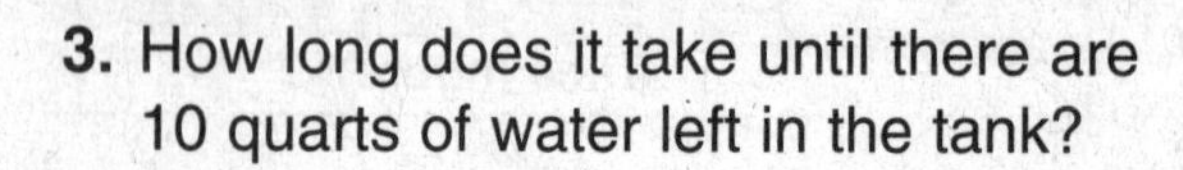

3. How long does it take until there are 10 quarts of water left in the tank?

____

4. How long does it take to drain the tank?

____

Use the graph to answer Exercises 5–7. The graph shows the distance, *d*, a hiker can hike in *h* hours. Choose the letter of the best answer.

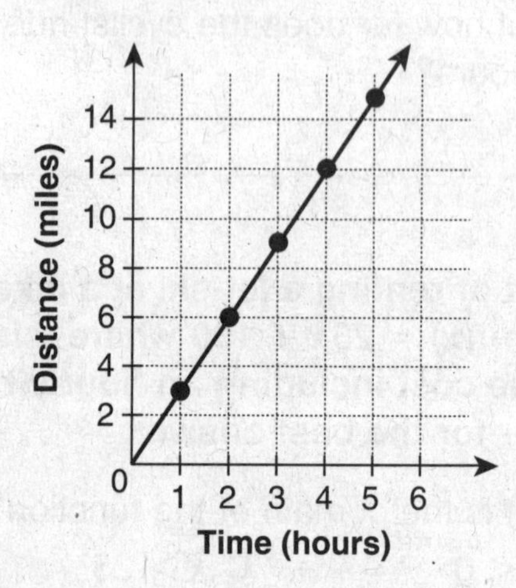

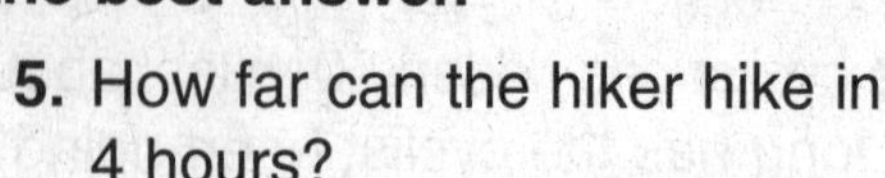

5. How far can the hiker hike in 4 hours?

A $1\frac{1}{3}$ mi **C** 8 mi

B 4 mi **D** 12 mi

6. How long does it take the hiker to hike 6 miles?

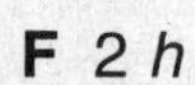

F 2 *h* **H** 4 *h*

G 3 *h* **J** 18 *h*

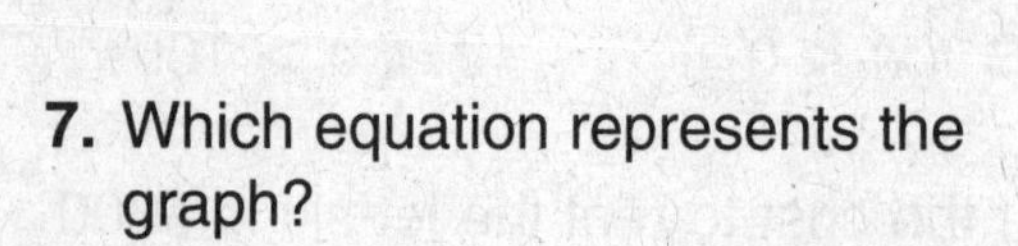

7. Which equation represents the graph?

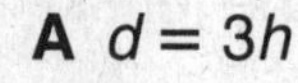

A $d = 3h$ **C** $d = h + 3$

B $d = \frac{1}{3}h$ **D** $d = h - 3$

Name ______________________ Date __________ Class __________

LESSON 3-6

Problem Solving

Arithmetic Sequences

Write the correct answer.

1. An English teacher gives her class 6 vocabulary words on Monday. Each day for the rest of the week she adds 3 more vocabulary words to the list. How many words are on the list on Friday?

____

2. A cab ride costs $1.50 plus $2.00 for each mile. What is the total cost of a 5-mile cab ride?

____

3. Rosie ran 8 laps around the track. Each week after that she ran 3 more laps than the week before. How many laps will she run around the track in the sixth week?

____

4. Lee has saved $85. Each week he uses his savings to buy a CD for $9. How much money will he have left after the fourth week?

____

Use the table to answer Exercises 5–7. The table shows the number of seats in each row of a theater. Choose the letter of the best answer.

Row	Number of Seats
1	35
2	43
3	51
4	59
5	67

5. The number of seats is an arithmetic sequence. What is the common difference?

A 6 **C** 9

B 8 **D** 35

6. If the sequence continues, how many seats will be in the next row?

F 68 **H** 75

G 73 **J** 76

7. If the sequence continues, how many seats will be in the tenth row?

A 80 **C** 134

B 107 **D** 147

8. A class is taking a field trip to the zoo. Admission for the class costs $50 plus $2 for each student to visit the special exhibits. Which function best describes the total cost for n students?

F $y = 50n - 2$

G $y = 50n + 2$

H $y = 50 - 2n$

J $y = 50 + 2n$

Name ______________________ Date ____________ Class ____________

LESSON 4-1

Problem Solving

Exponents

Write the correct answer.

1. The formula for the volume of a cube is $V = e^3$ where e is the length of a side of the cube. Find the volume of a cube with side length 6 cm.

____

2. The distance in feet traveled by a falling object is given by the formula $d = 16t^2$ where t is the time in seconds. Find the distance an object falls in 4 seconds.

____

3. The surface area of a cube can be found using the formula $S = 6e^2$ where e is the length of a side of the cube. Find the surface area of a cube with side length 6 cm.

____

4. John's father offers to pay him 1 cent for doing the dishes the first night, 2 cents for doing the dishes the second, 4 cents for the third, and so on, doubling each night. Write an expression using exponents for the amount John will get paid on the tenth night.

____

Use the table below for Exercises 5–7, which shows the number of e-mails forwarded at each level if each person continues a chain by forwarding an e-mail to 10 friends. Choose the letter for the best answer.

Forwarded E-mails

Level	E-mails forwarded
1	10
2	100
3	1000
4	10,000

5. How many e-mails were forwarded at level 5 alone?

A 5^{10} **C** 2^{10}

B 2^5 **D** 10^5

6. How many e-mails were forwarded at level 6 alone?

F 100,000 **H** 10,000,000

G 1,000,000 **J** 100,000,000

7. Forwarding chain e-mails can create problems for e-mail servers. Find out how many total e-mails have been forwarded after 6 levels.

A 1,111,110 **C** 1,000,000

B 6,000,000 **D** 100,000,000

Name ______________________ Date __________ Class __________

LESSON 4-2

Problem Solving

Look for a Pattern in Integer Exponents

Write the correct answer.

1. The weight of 10^7 dust particles is 1 gram. Evaluate 10^7.

____

2. The weight of one dust particle is 10^{-7} gram. Evaluate 10^{-7}.

____

3. As of 2001, only 10^6 rural homes in the United States had broadband Internet access. Evaluate 10^6.

____

4. Atomic clocks measure time in microseconds. A microsecond is 10^{-6} second. Evaluate 10^{-6}.

____

Choose the letter for the best answer.

5. The diameter of the nucleus of an atom is about 10^{-15} meter. Evaluate 10^{-15}.

A 0.0000000000001

B 0.00000000000001

C 0.0000000000000001

D 0.000000000000001

6. The diameter of the nucleus of an atom is 0.000001 nanometer. How many nanometers is the diameter of the nucleus of an atom?

F $(-10)^5$

G $(-10)^6$

H 10^{-6}

J 10^{-5}

7. A ruby-throated hummingbird weighs about 3^{-2} ounce. Evaluate 3^{-2}.

A -9

B -6

C $\frac{1}{9}$

D $\frac{1}{6}$

8. A ruby-throated hummingbird breathes 2×5^3 times per minute while at rest. Evaluate this amount.

F 1,000

G 250

H 125

J 30

Name ______________________________ Date ____________ Class ____________

LESSON 4-3

Problem Solving

Properties of Exponents

Write each answer as a power.

1. Cindy separated her fruit flies into equal groups. She estimates that there are 2^{10} fruit flies in each of 2^2 jars. How many fruit flies does Cindy have in all?

____

2. Suppose a researcher tests a new method of pasteurization on a strain of bacteria in his laboratory. If the bacteria are killed at a rate of 8^9 per sec, how many bacteria would be killed after 8^2 sec?

____

3. A satellite orbits the earth at about 13^4 km per hour. How long would it take to complete 24 orbits, which is a distance of about 13^5 km?

____

4. The side of a cube is 3^4 centimeters long. What is the volume of the cube? (Hint: $V = s^3$.)

____

Use the table to answer Exercises 5–6. The table describes the number of people involved at each level of a pyramid scheme. In a pyramid scheme each individual recruits so many others to participate who in turn recruit others, and so on. Choose the letter of the best answer.

Pyramid Scheme

Each person recruits 5 others.

Level	Total Number of People
1	5
2	5^2
3	5^3
4	5^4

5. Using exponents, how many people will be involved at level 6?

A 6^6 **C** 5^5

B 6^5 **D** 5^6

6. How many times more people will be involved at level 6 than at level 2?

F 5^4 **H** 5^5

G 5^3 **J** 5^6

7. There are 10^3 ways to make a 3-digit combination, but there are 10^6 ways to make a 6-digit combination. How many times more ways are there to make a 6-digit combination than a 3-digit combination?

A 5^{10} **C** 2^5

B 2^{10} **D** 10^3

8. After 3 hours, a bacteria colony has $(25^3)^3$ bacteria present. How many bacteria are in the colony?

F 25^1 **H** 25^9

G 25^6 **J** 25^{33}

Name ______________________ Date ____________ Class ____________

LESSON 4-4

Problem Solving

Scientific Notation

Write the correct answer.

1. In June 2001, the Intel Corporation announced that they could produce a silicon transistor that could switch on and off 1.5 trillion times a second. Express the speed of the transistor in scientific notation.

____

2. With this transistor, computers will be able to do 1×10^9 calculations in the time it takes to blink your eye. Express the number of calculations using standard notation.

____

3. The elements in this fast transistor are 20 nanometers long. A nanometer is one-billionth of a meter. Express the length of an element in the transistor in meters using scientific notation.

____

4. The length of the elements in the transistor can also be compared to the width of a human hair. The length of an element is 2×10^{-5} times smaller than the width of a human hair. Express 2×10^{-5} in standard notation.

____

Use the table to answer Exercises 5–9. Choose the best answer.

Distance From Earth To Stars
Light-Year = 5,880,000,000,000 mi.

Star	Constellation	Distance (light-years)
Sirius	Canis Major	8
Canopus	Carina	650
Alpha Centauri	Centaurus	4
Vega	Lyra	23

5. Express a light-year in miles using scientific notation.

A 58.8×10^{11} **C** 588×10^{10}

B 5.88×10^{12} **D** 5.88×10^{-13}

6. How many miles is it from Earth to the star Sirius?

F 4.705×10^{12} **H** 7.35×10^{12}

G 4.704×10^{13} **J** 7.35×10^{11}

7. How many miles is it from Earth to the star Canopus?

A 3.822×10^{15} **C** 3.822×10^{14}

B 1.230×10^{15} **D** 1.230×10^{14}

8. How many miles is it from Earth to the star Alpha Centauri?

F 2.352×10^{13} **H** 2.352×10^{14}

G 5.92×10^{13} **J** 5.92×10^{14}

9. How many miles is it from Earth to the star Vega?

A 6.11×10^{13} **C** 6.11×10^{14}

B 1.3524×10^{13} **D** 1.3524×10^{14}

Name ______________________________ Date ____________ Class ____________

LESSON 4-5

Problem Solving

Squares and Square Roots

Write the correct answer.

1. For college wrestling competitions, the NCAA requires that the wrestling mat be a square with an area of 1764 square feet. What is the length of each side of the wrestling mat?

____

2. For high school wrestling competitions, the wrestling mat must be a square with an area of 1444 square feet. What is the length of each side of the wrestling mat?

____

3. The Japanese art of origami requires folding square pieces of paper. Elena begins with a large sheet of square paper that is 169 square inches. How many squares can she cut out of the paper that are 4 inches on each side?

____

4. When the James family moved into a new house they had a square area rug that was 132 square feet. In their new house, there are three bedrooms. Bedroom one is 11 feet by 11 feet. Bedroom two is 10 feet by 12 feet and bedroom three is 13 feet by 13 feet. In which bedroom will the rug fit?

____

Choose the letter for the best answer.

5. A square picture frame measures 36 inches on each side. The actual wood trim is 2 inches wide. The photograph in the frame is surrounded by a bronze mat that measures 5 inches. What is the maximum area of the photograph?

A 841 sq. inches **B** 900 sq. inches
C 1156 sq. inches **D** 484 sq. inches

6. To create a square patchwork quilt wall hanging, square pieces of material are sewn together to form a larger square. Which number of smaller squares can be used to create a square patchwork quilt wall hanging?

F 35 squares **G** 64 squares
H 84 squares **J** 125 squares

7. A can of paint claims that one can will cover 400 square feet. If you painted a square with the can of paint, how long would it be on each side?

A 200 feet **B** 65 feet
C 25 feet **D** 20 feet

8. A box of tile contains 12 square tiles. If you tile the largest possible square area using whole tiles, how many tiles will you have left from the box?

F 9 **G** 6
H 3 **J** 0

Name ______________________ Date __________ Class __________

LESSON 4-6

Problem Solving

Estimating Square Roots

The distance to the horizon can be found using the formula $d = 112.88\sqrt{h}$ where d is the distance in kilometers and h is the number of kilometers from the ground. Round your answer to the nearest kilometer.

1. How far is it to the horizon when you are standing on the top of Mt. Everest, a height of 8.85 km?

____

2. Find the distance to the horizon from the top of Mt. McKinley, Alaska, a height of 6.194 km.

____

3. How far is it to the horizon if you are standing on the ground and your eyes are 2 m above the ground?

____

4. Mauna Kea is an extinct volcano on Hawaii that is about 4 km tall. You should be able to see the top of Mauna Kea when you are how far away?

____

You can find the approximate speed of a vehicle that leaves skid marks before it stops. The formulas $S = 5.5\sqrt{0.7L}$ and $S = 5.5\sqrt{0.8L}$, where S is the speed in miles per hour and L is the length of the skid marks in feet, will give the minimum and maximum speeds that the vehicle was traveling before the brakes were applied. Round to the nearest mile per hour.

5. A vehicle leaves a skid mark of 40 feet before stopping. What was the approximate speed of the vehicle before it stopped?

   A 25–35 mph **C** 29–31 mph
   B 28–32 mph **D** 68–70 mph

6. A vehicle leaves a skid mark of 100 feet before stopping. What was the approximate speed of the vehicle before it stopped?

   F 46–49 mph **H** 62–64 mph
   G 50–55 mph **J** 70–73 mph

7. A vehicle leaves a skid mark of 150 feet before stopping. What was the approximate speed of the vehicle before it stopped?

   A 50–55 mph **C** 55–70 mph
   B 53–58 mph **D** 56–60 mph

8. A vehicle leaves a skid mark of 200 feet before stopping. What was the approximate speed of the vehicle before it stopped?

   F 60–63 mph **G** 65–70 mph
   H 72–78 mph **J** 80–90 mph

Name ______________________ Date __________ Class __________

LESSON 4-7

Problem Solving

The Real Numbers

Write the correct answer.

1. Twin primes are prime numbers that differ by 2. Find an irrational number between twin primes 5 and 7.

____

2. Rounded to the nearest ten-thousandth, $\pi = 3.1416$. Find a rational number between 3 and π.

____

3. One famous irrational number is e. Rounded to the nearest ten-thousandth $e \approx 2.7823$. Find a rational number that is between 2 and e.

____

4. Perfect numbers are those that the divisors of the number sum to the number itself. The number 6 is a perfect number because $1 + 2 + 3 = 6$. The number 28 is also a perfect number. Find an irrational number between 6 and 28.

____

Choose the letter for the best answer.

5. Which is a rational number?
   - **A** the length of a side of a square with area 2 cm^2
   - **B** the length of a side of a square with area 4 cm^2
   - **C** a non-terminating decimal
   - **D** the square root of a prime number

6. Which is an irrational number?
   - **F** a number that can be expressed as a fraction
   - **G** the length of a side of a square with area 4 cm^2
   - **H** the length of a side of a square with area 2 cm^2
   - **J** the square root of a negative number

7. Which is an integer?
   - **A** the number half-way between 6 and 7
   - **B** the average rainfall for the week if it rained 0.5 in., 2.3 in., 0 in., 0 in., 0 in., 0.2 in., 0.75 in. during the week
   - **C** the money in an account if the balance was $213.00 and $21.87 was deposited
   - **D** the net yardage after plays that resulted in a 15 yard loss, 10 yard gain, 6 yard gain and 5 yard loss

8. Which is a whole number?
   - **F** the number half-way between 6 and 7
   - **G** the total amount of sugar in a recipe that calls for $\frac{1}{4}$ cup of brown sugar and $\frac{3}{4}$ cup of granulated sugar
   - **H** the money in an account if the balance was $213.00 and $21.87 was deposited
   - **J** the net yardage after plays that resulted in a 15 yard loss, 10 yard gain, 6 yard gain and 5 yard loss

Name ______________________ Date __________ Class __________

LESSON 4-8

Problem Solving

The Pythagorean Theorem

Write the correct answer. Round to the nearest tenth.

1. A utility pole 10 m high is supported by two guy wires. Each guy wire is anchored 3 m from the base of the pole. How many meters of wire are needed for the guy wires?

____

2. A 12 foot-ladder is resting against a wall. The base of the ladder is 2.5 feet from the base of the wall. How high up the wall will the ladder reach?

____

3. The base-path of a baseball diamond form a square. If it is 90 ft from home to first, how far does the catcher have to throw to catch someone stealing second base?

____

4. A football field is 100 yards with 10 yards at each end for the end zones. The field is 45 yards wide. Find the length of the diagonal of the entire field, including the end zones.

____

Choose the letter for the best answer.

5. The frame of a kite is made from two strips of wood, one 27 inches long, and one 18 inches long. What is the perimeter of the kite? Round to the nearest tenth.

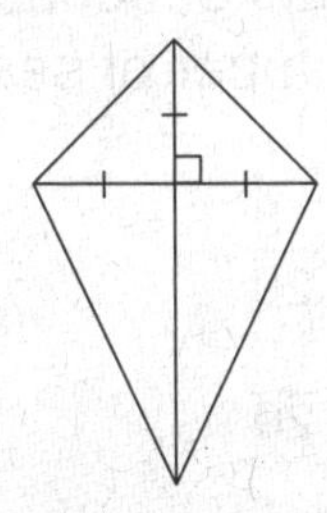

A 18.8 in. **C** 65.7 in.
B 32.8 in. **D** 131.2 in.

6. The glass for a picture window is 8 feet wide. The door it must pass through is 3 feet wide. How tall must the door be for the glass to pass through the door? Round to the nearest tenth.

F 3.3 ft **H** 7.4 ft
G 6.7ft **J** 8.5 ft

7. A television screen measures approximately 15.5 in. high and 19.5 in. wide. A television is advertised by giving the approximate length of the diagonal of its screen. How should this television be advertised?

A 25 in. **C** 12 in.
B 21 in. **D** 6 in.

8. To meet federal guidelines, a wheelchair ramp that is constructed to rise 1 foot off the ground must extend 12 feet along the ground. How long will the ramp be? Round to the nearest tenth.

F 11.9 ft **H** 13.2 ft
G 12.0 ft **J** 15.0 ft

Name ______________________ Date __________ Class __________

LESSON 5-1

Problem Solving

Ratios and Proportions

A medicine for dogs indicates that the medicine should be administered in the ratio 0.5 tsp per 5 lb, based on the weight of the dog. Write the correct answer.

1. Jaime has a 60 lb dog. She plans to give the dog 12 teaspoons of medicine. Is she administering the medicine correctly?

   ____

2. Jaime also has a 15 lb puppy. She plans to give the puppy 1.5 teaspoons of medicine. Is she administering the medicine correctly?

   ____

Sports statistics can be given as ratios. Find the ratios for the given statistics. Reduce each ratio.

3. In 69 games, Darrel Armstrong of the Orlando Magic had 136 steals and 144 turnovers. What is his steals per turnover ratio?

   ____

4. In 69 games, Ben Wallace of the Detroit Pistons blocked 234 shots. What is his blocks per game ratio?

   ____

Choose the letter for the best answer.

5. There are 675 students and 30 teachers in the middle school. What is the ratio of teachers to students?

   A $\frac{45}{2}$ **C** $\frac{1}{27}$

   B $\frac{2}{45}$ **D** $\frac{27}{1}$

6. In a science experiment, out of a sample of seeds, 13 sprouted and 7 didn't. What is the ratio of seeds that sprouted to the number of seeds planted?

   F $\frac{13}{7}$ **H** $\frac{13}{20}$

   G $\frac{7}{13}$ **J** $\frac{7}{20}$

7. Many Internet services advertise their customer to modem ratio. One company advertises a 10 to 1 customer to modem ratio. Find a ratio that is equivalent to $\frac{10}{1}$.

   A $\frac{40}{4}$ **C** $\frac{400}{4}$

   B $\frac{2}{20}$ **D** $\frac{50}{10}$

8. A molecule of sulfuric acid contains 2 atoms of hydrogen to every 4 atoms of oxygen. Which combination of hydrogen and oxygen atoms could be sulfuric acid?

   F 4 atoms of hydrogen and 6 atoms of oxygen

   G 6 atoms of hydrogen and 10 atoms of oxygen

   H 6 atoms of hydrogen and 12 atoms of oxygen

   J 16 atoms of hydrogen and 8 atoms of oxygen

Name ______________________ Date __________ Class __________

LESSON 5-2

Problem Solving

Ratios, Rates, and Unit Rates

Scientists have researched the ratio of brain weight to body size in different animals. The results are in the table below.

Animal	$\frac{\text{Brain Weight}}{\text{Body Weight}}$
Cat	$\frac{1}{100}$
Dog	$\frac{1}{125}$
Elephant	$\frac{1}{560}$
Hippo	$\frac{1}{2789}$
Horse	$\frac{1}{600}$
Human	$\frac{1}{40}$
Small birds	$\frac{1}{12}$

1. Order the animals by their brain weight to body weight ratio, from smallest to largest.

____

____

2. It has been hypothesized that the higher the brain weight to body weight ratio, the more intelligent the animal is. By this measure, which animals listed are the most intelligent?

____

3. Name two sets of animals that have approximately the same brain weight to body weight ratio.

____

Find the unit rate. Round to the nearest hundredth.

4. A 64-ounce bottle of apple juice costs $1.35.
   - **A** $0.01/oz
   - **B** $0.02/oz
   - **C** $0.47/oz
   - **D** $47.4/oz

5. Find the unit rate for a 2 lb package of hamburger that costs $3.45.
   - **F** $0.58/lb
   - **G** $1.25/b
   - **H** $1.73/lb
   - **J** $2.28/b

6. 12 slices of pizza cost $9.00.
   - **A** $0.45/slice
   - **B** $0.50/slice
   - **C** $0.75/slice
   - **D** $1.33/slice

7. John is selling 5 comic books for $6.00.
   - **F** $0.83/book
   - **G** $1.20/book
   - **H** $1.02/book
   - **J** $1.45/book

8. There are 64 beats in 4 measures of music.
   - **A** 16 beats/measure
   - **B** 12 beats/measure
   - **C** 4 beats/measure
   - **D** 0.06 beats/measure

9. The average price of a 30 second commercial for the 2002 Super Bowl was $1,900,000.
   - **F** $120.82/sec
   - **G** $1,242.50/sec
   - **H** $5,839.02/sec
   - **J** $63,333.33/sec

Name ______________________ Date __________ Class __________

LESSON 5-3

Problem Solving

Dimensional Analysis

Use the following: 1 mile = 1.609 km; 1 kg = 2.2046 lb. Round to the nearest tenth.

1. Worker bees travel up to 14 km to find pollen and nectar. How far will a worker bee travel in miles?

   ____

2. Worker bees can travel at 24 km/h. How fast can the worker bee travel in miles per hour?

   ____

3. The average hippopotamus weighs 1800 kg. How many pounds does the average hippopotamus weigh?

   ____

4. At the age of 45, an elephant grows teeth, each weighing 4 kg. How many pounds do these teeth weigh?

   ____

Paraceratherium was the biggest land mammal there has ever been. It lived about 35 million years ago and was 8 m tall and 11 m long. It looked like a gigantic rhinoceros but had a long neck like a giraffe. 1 foot = 0.3048 meters. Round to the nearest tenth.

5. How tall was the paraceratherium in feet?

   ____

6. How long was the paraceratherium in feet?

   ____

Round to the nearest tenth. Choose the letter for the best answer.

7. The fastest sporting animal is the racing pigeon that flies up to 110 mi an hour. How fast is the racing pigeon in feet each second?

   A 75.0 ft/s **C** 543.2 ft/s
   B 161.3 ft/s **D** 9,680 ft/s

8. The longest gloved fight between two Americans lasted for more than seven hours before being declared a draw. How many seconds did the fight last?

   F 127 s **H** 420 s
   G 385 s **J** 25,200 s

9. The average person falls asleep in seven minutes. How many seconds does it take the average person to fall asleep?

   A 127 s **C** 420 s
   B 385 s **D** 25,200 s

10. The brain of an average adult male weighs 55 oz. How many pounds does the average male brain weigh?

    F 3.4 lb **H** 13.8 lb
    G 5.8 lb **J** 880 lb

Name ________________________ Date __________ Class __________

LESSON 5-4

Problem Solving

Solving Proportions

Use the ratios in the table to answer each question. Round to the nearest tenth.

Body Part	Body Part / Height
Femur	$\frac{1}{4}$
Tibia	$\frac{1}{5}$
Hand span	$\frac{2}{17}$
Arm span	$\frac{1}{1}$
Head circumference	$\frac{1}{3}$

1. Which body part is the same length as the person's height?

____

2. If a person's tibia is 13 inches, how tall would you expect the person to be?

____

3. If a person's hand span is 8.5 inches, about how tall would you expect the person to be?

____

4. If a femur is 18 inches long, how many feet tall would you expect the person to be?

____

5. What would you expect the head circumference to be of a person who is 5.5 feet tall?

____

6. What would you expect the hand span to be of a person who is 5 feet tall?

____

Choose the letter for the best answer.

7. Five milliliters of a children's medicine contains 400 mg of the drug amoxicillin. How many mg of amoxicillin does 25 mL contain?

A 0.3 mg **C** 2000 mg
B 80 mg **D** 2500 mg

8. Vladimir Radmanovic of the Seattle Supersonics makes, on average, about 2 three-pointers for every 5 he shoots. If he attempts 10 three-pointers in a game, how many would you expect him to make?

F 4 **H** 8
G 5 **J** 25

9. In 2002, a 30-second commercial during the Super Bowl cost an average of $1,900,000. At this rate, how much would a 45-second commercial cost?

A $1,266,666 **C** $3,500,000
B $2,850,000 **D** $4,000,000

10. A medicine for dogs indicates that the medicine should be administered in the ratio 2 teaspoons per 5 lb, based on the weight of the dog. How much should be given to a 70 lb dog?

F 5 teaspoons **H** 14 teaspoons
G 12 teaspoons **J** 28 teaspoons

Name ______________________ Date ____________ Class ____________

LESSON 5-5

Problem Solving

Similar Figures

Write the correct answer.

1. Until 1929, United States currency measured 3.13 in. by 7.42 in. The current size is 2.61 in. by 6.14 in. Are the bills similar?

____

2. Owen has a 3 in. by 5 in. photograph. He wants to make it as large as he can to fit in a 10 in. by 12.5 in. ad. What scale factor will he use? What will be the new size?

____

3. A painting is 15 cm long and 8 cm wide. In a reproduction that is similar to the original painting, the length is 36 cm. How wide is the reproduction?

____

4. The two shortest sides of a right triangle are 10 in. and 24 in. long. What is the length of the shortest side of a similar right triangle whose two longest sides are 36 in. and 39 in.?

____

The scale on a map is 1 inch = 40 miles. Round to the nearest mile.

5. On the map, it is 5.75 inches from Orlando to Miami. How many miles is it from Orlando to Miami?

A 46 miles
B 175 miles
C 230 miles
D 340 miles

6. On the map it is $18\frac{1}{8}$ inches from Norfolk, VA, to Indianapolis, IN. How many miles is it from Norfolk to Indianapolis?

F 58 miles
G 725 miles
H 800 miles
J 1025 miles

7. It is 185 miles from Chicago to Indianapolis. On the map it is 2.5 inches from Indianapolis to Terra Haute, IN. How far is it from Chicago to Terra Haute going through Indianapolis?

A 100 miles
B 285 miles
C 430 miles
D 7500 miles

8. On the map, it is 7.5 inches from Chicago to Cincinnati. Traveling at 65 mi/h, how long will it take to drive from Chicago to Cincinnati? Round to the nearest tenth of an hour.

F 4.6 hours
G 5.2 hours
H 8.7 hours
J 12.0 hours

Name ______________________________ Date ____________ Class ____________

LESSON 5-6

Problem Solving

Dilations

Write the correct answer.

1. When you enlarge something on a photocopy machine, is the image a dilation?

____

2. When you make a photocopy that is the same size, is the image a dilation? If so, what is the scale factor?

____

3. In the movie *Honey, I Blew Up the Kid*, a two-year-old-boy is enlarged to a height of 112 feet. If the average height of a two-year old boy is 3 feet, what is the scale factor of this enlargement?

____

4. In the movie *Honey, I Shrunk the Kids*, an inventor shrinks his kids by a scale factor of about $\frac{1}{240}$. If his kids were about 5 feet tall, how many inches tall were they after they were shrunk?

____

Use the coordinate plane for Exercises 5–6. Round to the nearest tenth. Choose the letter for the best answer.

5. What will be the coordinates of A', B' and C' after $\triangle ABC$ is dilated by a factor of 5?

A $A'(5, 10)$, $B'(7, 8)$, $C'(3, 8)$

B $A'(0, 25)$, $B'(10, 15)$, $C'(-10, 15)$

C $A'(0, 5)$, $B'(10, 3)$, $C'(-10, 3)$

D $A'(0, 15)$, $B'(2, 15)$, $C'(-2, 15)$

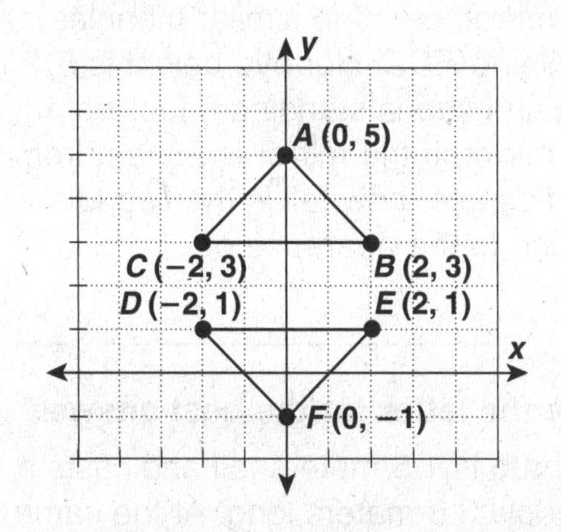

6. What will be the coordinates of D', E' and F' after $\triangle DEF$ is dilated by a factor of 5?

F $D'(-10, 5)$, $E'(10, 5)$, $F'(0, -5)$

G $D'(10, 5)$, $E'(5, 5)$, $F'(0, 5)$

H $D'(10, 0)$, $E'(5, 0)$, $F'(-5, 0)$

J $D'(10, 5)$, $E'(5, 5)$, $F'(-10, 0)$

7. The projection of a movie onto a screen is a dilation. The universally accepted film size for movies has a width of 35 mm. If a movie screen is 12 m wide, what is the dilation factor?

A 420

B 0.3

C 342.9

D 2916.7

Name ______________________ Date ____________ Class ____________

LESSON 5-7

Problem Solving

Indirect Measurement

Write the correct answer.

1. Celine wants to know the width of the pond. She drew the diagram shown below and labeled it with the measurements she made. How wide is the pond?

____

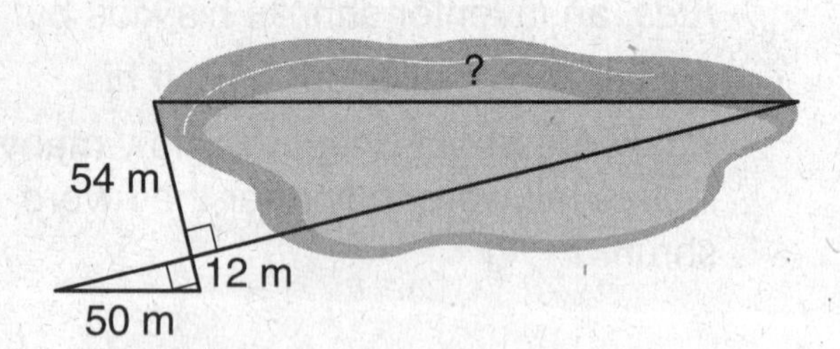

2. Vince wants to know the distance across the canyon. He drew the diagram and labeled it with the measurements he made. What is the distance across the canyon?

____

?
25 ft
5 ft
17 ft

3. Paula places a mirror between herself and a flagpole. She stands so she can see the top of the flagpole in the mirror, creating similar triangles *ABC* and *EDC*. Her eye height is 5 feet and she is standing 6 feet from the mirror. If the mirror is 25 feet from the flagpole, how tall is the flagpole? Round to the nearest foot.

____

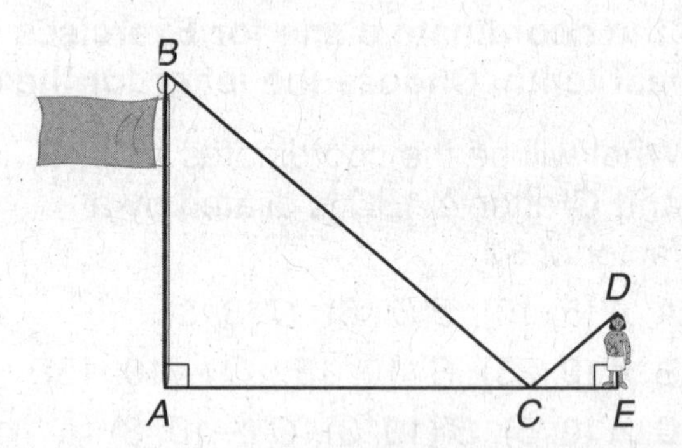

Choose the letter for the best answer.

4. A shrub is 1.5 meters tall and casts a shadow 3.5 meters long. At the same time, a radio tower casts a shadow 98 meters long. How tall is the radio tower?

A 33 m

B 42 m

C 147 m

D 329 m

5. Kim is 56 inches tall. His friend is 42 inches tall. Kim's shadow is 24 inches long. How long is his friend's shadow at the same time?

F 18 in.

G 32 in.

H 38 in.

J 98 in.

Name ______________________ Date __________, Class __________

LESSON 5-8

Problem Solving

Scale Drawings and Scale Models

Round to the nearest tenth. Write the correct answer.

1. The Statue of Liberty is approximately 305 feet tall. A scale model of the Statue of Liberty is 5 inches tall. The scale of the model is 1 in. : ____ ft.

____

2. The right arm of the Statue of Liberty is 42 feet long. How long is the right arm of the Statue of Liberty model given in Exercise 1?

____

3. The diameter of an atom is 10^{-9} cm. If a scale drawing of an atom has a diameter of 10 cm, the scale of the drawing is 1 cm : ____ cm.

____

4. The diameter of the nucleus of an atom is 10^{-13} cm. If a scale drawing of the nucleus of an atom has a diameter of 1 cm, the scale of the drawing is 1 cm : ____ cm.

____

The toy car in Exercises 5-6 has a scale of $\frac{1}{40}$. Choose the letter for the best answer.

5. The diameter of the steering wheel of the actual car is 15 inches. What is the diameter of the toy car's steering wheel?

A $\frac{3}{8}$ in.
B $\frac{1}{2}$ in.
C $1\frac{1}{2}$ in.
D $2\frac{2}{3}$ in.

6. The diameter of the toy car's tire is $\frac{5}{8}$ in. What is the diameter of the tire of the actual car?

F $12\frac{1}{2}$ in.
G 16 in.
H 25 in.
J 64 in.

7. On a scale drawing, a 14 ft room is pictured as 3.5 inches. What is the scale of the drawing?

A $\frac{1}{4}$ in.
B $\frac{1}{2}$ in.
C $\frac{1}{4}$:14
D 1:56

8. On a scale drawing, $\frac{1}{2}$ inch = 1 foot. A room is pictured as 7.5 inches by 6 inches. How many square yards of carpet are needed for the room?

F 5 yd^2
G 20 yd^2
H 45 yd^2
J 90 yd^2

9. On a scale drawing of a computer component, $\frac{1}{4}$ in. = 4 in. On the drawing, a piece is $\frac{3}{8}$ in. long. How long is the actual piece?

A 1.5 in.
B 3 in.
C 6 in.
D 7.5 in.

10. A scale drawing has a $\frac{1}{4}$ inch scale. The width of a 12 foot room is going to be increased by 4 feet. How much wider will the room be on the drawing?

F $\frac{1}{4}$ in.
G $\frac{1}{2}$ in.
H 1 in.
J 4 in.

Name ______________________ Date ____________ Class ____________

LESSON 6-1

Problem Solving

Relating Decimals, Fractions, and Percents

The table shows the ratio of brain weight to body size in different animals. Use the table for Exercises 1–3. Write the correct answer.

1. Complete the table to show the percent of each animal's body weight that is brain weight. Round to the nearest hundredth.

Animal	$\frac{\text{Brain Weight}}{\text{Body Weight}}$	Percent
Mouse	$\frac{1}{40}$	
Cat	$\frac{1}{100}$	
Dog	$\frac{1}{125}$	
Horse	$\frac{1}{600}$	
Elephant	$\frac{1}{560}$	

2. Which animal has a greater brain weight to body size ratio, a dog or an elephant?

____

3. List the animals from least to greatest brain weight to body size ratio.

____

The table shows the number of wins and losses of the top teams in the National Football Conference from 2004. Choose the letter of the best answer. Round to the nearest tenth.

Team	Wins	Losses
Philadelphia Eagles	13	3
Green Bay Packers	10	6
Atlanta Falcons	11	5
Seattle Seahawks	9	7

4. What percent of games did the Green Bay Packers win?

A 10% **C** 37.5%

B 60% **D** 62.5%

5. Which decimal is equivalent to the percent of games the Seattle Seahawks won?

F 0.05625 **H** 5.625

G 0.5625 **J** 56.25

6. Which team listed had the highest percentage of wins?

A Philadelphia Eagles

B Green Bay Packers

C Atlanta Falcons

D Seattle Seahawks

Name ______________________ Date __________ Class __________

LESSON 6-2

Problem Solving

Estimate with Percents

Write an estimate.

1. A store requires you to pay 15% up front on special orders. If you plan to special order items worth $74.86, estimate how much you will have to pay up front.

____

2. A store is offering 25% off of everything in the store. Estimate how much you will save on a jacket that is normally $58.99.

____

3. A certain kind of investment requires that you pay a 10% penalty on any money you remove from the investment in the first 7 years. If you take $228 out of the investment, estimate how much of a penalty you will have to pay.

____

4. John notices that about 18% of the earnings from his job go to taxes. If he works 14 hours at $6.25 an hour, about how much of his check will go for taxes?

____

Choose the letter for the best estimate.

5. In its first week, an infant can lose up to 10% of its body weight in the first few days of life. Which is a good estimate of how many ounces a 5 lb 13 oz baby might lose in the first week of life?

A 0.6 oz **C** 18 oz
B 9 oz **D** 22 oz

6. A CD on sale costs $12.89. Sales tax is 4.75%. Which is the best estimate of the total cost of the CD?

F $13.30 **H** $14.20
G $13.55 **J** $14.50

7. In a class election, Pedro received 52% of the votes. There were 274 students who voted in the election. Which is the best estimate of the number of students who voted for Pedro?

A 70 students **C** 125 students
B 100 students **D** 140 students

8. Mel's family went out for breakfast. The bill was $24.25 plus 5.2% sales tax. Mel wants to leave a 20% tip. Which is the best estimate of the total bill?

F $25.45 **H** $30.25
G $29.25 **J** $32.25

Name ______________________ Date ____________ Class ____________

LESSON 6-3

Problem Solving

Finding Percents

Write the correct answer.

1. Florida State University in Tallahassee, Florida has 29,820 students. Approximately 60% of the students are women. How many of the students are women?

____

2. The yearly cost of tuition, room and board at Florida State University for a Florida resident is $10,064. If tuition is $3,208 a year, what percent of the yearly cost is tuition? Round to the nearest tenth of a percent.

____

3. The yearly cost of tuition, room and board at Florida State University for a non-Florida resident is $23,196. If tuition is $16,340 a year, what percent of the yearly cost is tuition for a non-resident? Round to the nearest tenth of a percent.

____

4. Approximately 65% of the students who apply to Florida State University are accepted. If 15,000 students apply to Florida State University, how many would you expect to be accepted?

____

The top four NBA field goal shooters for the 2003–2004 regular season are given in the table below. Choose the letter for the best answer.

NBA Field Goal Leaders 2003–2004 Season

Player	Attempts	Made	Percent
Shaquille O'Neal	948	554	
Donyell Marshall	604		56.6
Elton Brand	950	348	
Dale Davis	913		53.5

5. What percent of field goals did Shaquille O'Neal make? Round to the nearest tenth of a percent.

A 0.6% **C** 58.4%

B 1.71% **D** 59.2%

6. How many field goals did Donyell Marshall make in the 2003–2004 regular season?

F 38 **H** 295

G 114 **J** 342

7. What percent of field goals did Dale Davis make? Round to the nearest tenth of a percent.

A 1.87% **C** 51.9%

B 50% **D** 53.5%

8. How many field goals did Antawr Jamison make in the 2003–2004 regular season?

F 274 **H** 457

G 378 **J** 488

Name ______________________ Date __________ Class __________

LESSON 6-4

Problem Solving

Finding a Number When the Percent is Known

Write the correct answer.

1. The two longest running Broadway shows are *Cats* and *A Chorus Line*. *A Chorus Line* had 6137, or about 82% of the number of performances that *Cats* had. How many performances of *Cats* were there?

____

2. *Titanic* and *Star Wars* have made the most money at the box office. *Star Wars* made about 76.7% of the money that *Titanic* made at the box office. If *Star Wars* made about $461 million, how much did *Titanic* make? Round to the nearest million dollars.

____

Use the table below. Round to the nearest tenth of a percent.

3. What percent of students are in Pre-K through 8th grade?

____

4. What percent of students are in grades 9–12?

____

Public Elementary and Secondary School Enrollment, 2001

Grades	Population (in thousands)
Pre-K through grade 8	33,952
Grades 9–12	13,736
Total	47,688

Choose the letter for the best answer.

5. In 2000, women earned about 72.2% of what men did. If the average woman's weekly earnings was $491 in 2000, what was the average man's weekly earnings? Round to the nearest dollar.

A $355 **C** $680
B $542 **D** $725

6. The highest elevation in North America is Mt. McKinley at 20,320 ft. The highest elevation in Australia is Mt. Kosciusko, which is about 36% of the height of Mt. McKinley. What is the highest elevation in Australia? Round to the nearest foot.

F 5480 ft **H** 12,825 ft
G 7315 ft **J** 56,444 ft

7. The Gulf of Mexico has an average depth of 4,874 ft. This is about 36.2% of the average depth of the Pacific Ocean. What is the average depth of the Pacific Ocean? Round to the nearest foot.

A 1764 ft **C** 10,280 ft
B 5843 ft **D** 13,464 ft

8. Karl Malone is the NBA lifetime leader in free throws. He attempted 11,703 and made 8,636. What percent did he make? Round to the nearest tenth of a percent.

F 1.4% **H** 73.8%
G 58.6% **J** 135.6%

Name ______________________ Date ___________ Class ___________

LESSON 6-5

Problem Solving

Percent Increase and Decrease

Use the table below. Write the correct answer.

1. What is the percent increase in the population of Las Vegas, NV from 1990 to 2000? Round to the nearest tenth of a percent.

____

2. What is the percent increase in the population of Naples, FL from 1990 to 2000? Round to the nearest tenth of a percent.

____

Fastest Growing Metropolitan Areas, 1990–2000

Metropolitan Area	Population		Percent Increase
	1990	2000	
Las Vegas, NV	852,737	1,563,282	
Naples, FL	152,099	251,377	
Yuma, AZ	106,895		49.7%
McAllen-Edinburg-Mission, TX	383,545		48.5%

3. What was the 2000 population of Yuma, AZ to the nearest whole number?

____

4. What was the 2000 population of McAllen-Edinburg-Mission, TX metropolitan area to the nearest whole number?

____

For exercises 5–7, round to the nearest tenth. Choose the letter for the best answer.

5. The amount of money spent on automotive advertising in 2000 was 4.4% lower than in 1999. If the 1999 spending was $1812.3 million, what was the 2000 spending?

A $79.7 million **C** $1892 million

B $1732.6 million **D** $1923.5 million

6. In 1967, a 30-second Super Bowl commercial cost $42,000. In 2000, a 30-second commercial cost $1,900,000. What was the percent increase in the cost?

F 1.7% **H** 442.4%

G 44.2% **J** 4423.8%

7. In 1896 Thomas Burke of the U.S. won the 100-meter dash at the Summer Olympics with a time of 12.00 seconds. In 2004, Justin Gatlin of the U.S. won with a time of 9.85 seconds. What was the percent decrease in the winning time?

A 2.15% **C** 21.8%

B 17.9% **D** 45.1%

8. In 1928 Elizabeth Robinson won the 100-meter dash with a time of 12.20 seconds. In 2004, Yuliya Nesterenko won with a time that was about 10.4% less than Robinson's winning time. What was Nesterenko's time, rounded to the nearest hundredth?

F 9.83 seconds **H** 12.16 seconds

G 10.93 seconds **J** 13.47 seconds

Name ______________________ Date __________ Class __________

LESSON 6-6

Problem Solving

Applications of Percents

Write the correct answer.

1. The sales tax rate for a community is 6.75%. If you purchase an item for $500, how much will you pay in sales tax?

____

2. A community is considering increasing the sales tax rate 0.5% to fund a new sports arena. If the tax rate is raised, how much more will you pay in sales tax on $500?

____

3. Trent earned $28,500 last year. He paid $8,265 for rent. What percent of his earnings did Trent pay for rent?

____

4. Julie has been offered two jobs. The first pays $400 per week. The second job pays $175 per week plus 15% commission on her sales. How much will she have to sell in order for the second job to pay as much as the first?

____

Choose the letter for the best answer. Round to the nearest cent.

5. Clay earned $2,600 last month. He paid $234 for entertainment. What percent of his earnings did Clay pay in entertainment expenses?

   A 9%

   B 11%

   C 30%

   D 90%

6. Susan's parents have offered to help her pay for a new computer. They will pay 30% and Susan will pay 70% of the cost of a new computer. Susan has saved $550 for a new computer. With her parents help, how expensive of a computer can she afford?

   F $165.00 **H** $1650.00

   G $785.71 **J** $1833.33

7. Kellen's bill at a restaurant before tax and tip is $22.00. If tax is 5.25% and he wants to leave 15% of the bill including the tax for a tip, how much will he spend in total?

   A $22.17 **C** $26.63

   B $26.46 **D** $27.82

8. The 8th grade class is trying to raise money for a field trip. They need to raise $600 and the fundraiser they have chosen will give them 20% of the amount that they sell. How much do they need to sell to raise the money for the field trip?

   F $120.00 **H** $3000.00

   G $857.14 **J** $3200.00

Name ______________________ Date __________ Class __________

LESSON 6-7

Problem Solving

Simple Interest

Write the correct answer.

1. Joanna's parents agree to loan her the money for a car. They will loan her $5,000 for 5 years at 5% simple interest. How much will Joanna pay in interest to her parents?

____

2. How much money will Joanna have spent in total on her car with the loan described in exercise 1?

____

3. A bank offers simple interest on a certificate of deposit. Jaime invests $500 and after one year earns $40 in interest. What was the interest rate on the certificate of deposit?

____

4. How long will Howard have to leave $5000 in the bank to earn $250 in simple interest at 2%?

____

Jan and Stewart Jones plan to borrow $20,000 for a new car. They are trying to decide whether to take out a 4-year or 5-year simple interest loan. The 4-year loan has an interest rate of 6% and the 5-year loan has an interest rate of 6.25%. Choose the letter for the best answer.

5. How much will they pay in interest on the 4-year loan?
   A $4500 **C** $5000
   B $4800 **D** $5200

6. How much will they repay with the 4-year loan?
   F $24,500 **H** $25,000
   G $24,800 **J** $25,200

7. How much will they pay in interest on the 5-year loan?
   A $5000 **C** $6250
   B $6000 **D** $6500

8. How much will they repay with the 5-year loan?
   F $25,000 **H** $26,250
   G $26,000 **J** $26,500

9. How much more interest will they pay with the 5-year loan?
   A $1000
   B $1450
   C $1500
   D $2000

10. If the Stewarts can get a 5-year loan with 5.75% simple interest, which of the loans is the best deal?
   F 4 year, 6%
   G 5 year, 5.75%
   H 5 year, 6.25%
   J Cannot be determined

Name ______________________ Date __________ Class __________

LESSON 7-1

Problem Solving

Points, Lines, Planes, and Angles

Use the flag of the Bahamas to solve the problems.

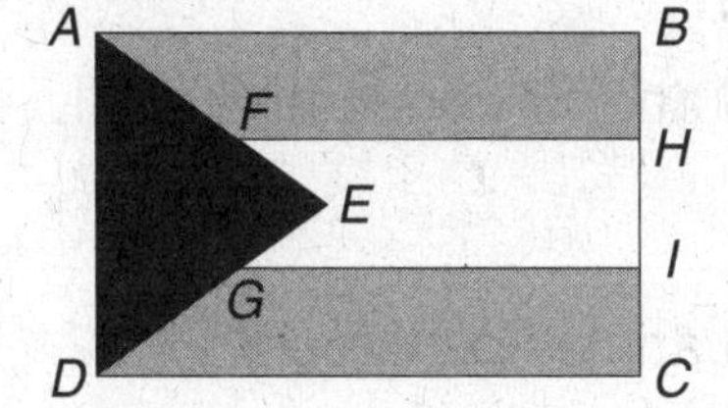

1. Name four points in the flag.

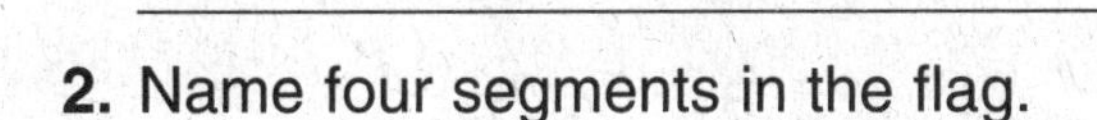

2. Name four segments in the flag.

____

3. Name a right angle in the flag.

____

4. Name two acute angles in the flag.

____

5. Name a pair of complementary angles in the flag.

____

6. Name a pair of supplementary angles in the flag.

____

The diagram illustrates a ray of light being reflected off a mirror. The angle of incidence is congruent to the angle of reflection. Choose the letter for the best answer.

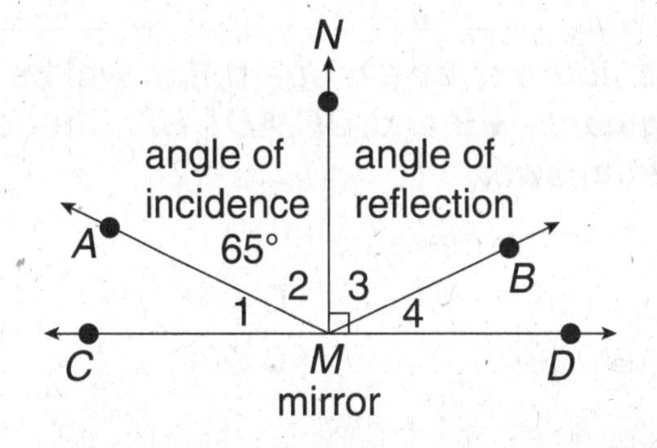

7. Name two rays in the diagram.

   A $\overline{AM}$, $\overline{MB}$ **C** $\overrightarrow{MA}$, $\overrightarrow{MB}$

   B $\overrightarrow{MA}$, $\overrightarrow{BM}$ **D** $\overrightarrow{MA}$, $\overrightarrow{MB}$

8. Name a pair of complementary angles.

   F $\angle NMB$, $\angle BMD$ **H** $\angle CMA$, $\angle AMD$

   G $\angle AMN$, $\angle NMB$ **J** $\angle CMA$, $\angle DMB$

9. Which angle is congruent to $\angle 2$?

   A $\angle 1$ **C** $\angle 3$

   B $\angle 4$ **D** none

10. Find the measure of $\angle 4$.

    F 65° **H** 25°

    G 35° **J** 90°

11. Find the measure of $\angle 1$.

    A 65° **C** 25°

    B 35° **D** 90°

12. Find the measure of $\angle 3$.

    F 90° **H** 35°

    G 45° **J** 65°

Name ______________________ Date ___________ Class ___________

LESSON **7-2**

Problem Solving

Parallel and Perpendicular Lines

The figure shows the layout of parking spaces in a parking lot.
$\overline{AB} \parallel \overline{CD} \parallel \overline{EF}$

1. Name all angles congruent to $\angle 1$.

____

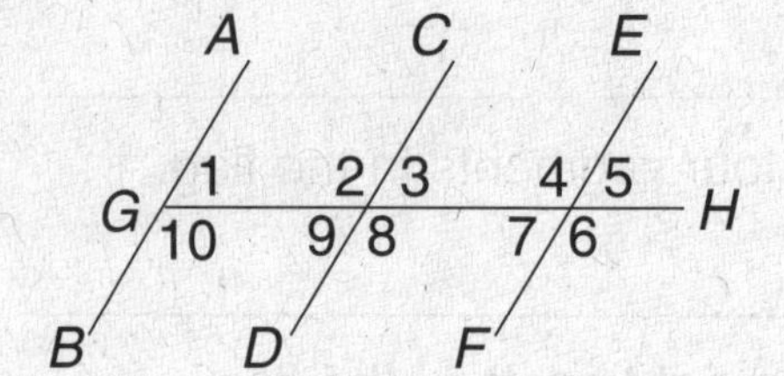

2. Name all angles congruent to $\angle 2$.

____

3. Name a pair of supplementary angles.

____

4. If $m\angle 1 = 75°$, find the measures of the other angles.

____

____

____

5. Name a pair of vertical angles.

____

6. If $m\angle 1 = 90°$, then $\overline{GH}$ is perpendicular to

_______________________.

The figure shows a board that will be cut along parallel segments *GB* and *CF*. $\overline{AD} \parallel \overline{HE}$. Choose the letter for the best answer.

7. Find the measure of $\angle 1$.

A 45° **C** 60°
B 120° **D** 90°

8. Find the measure of $\angle 2$.

F 30° **H** 60°
G 120° **J** 90°

9. Find the measure of $\angle 3$.

A 30° **C** 60°
B 120° **D** 90°

10. Find the measure of $\angle 4$.

F 45° **H** 60°
G 120° **J** 90°

11. Find the measure of $\angle 5$.

A 30° **C** 60°
B 120° **D** 90°

12. Find the measure of $\angle 6$.

F 30° **H** 60°
G 120° **J** 90°

13. Find the measure of $\angle 7$.

A 45° **C** 60°
B 120° **D** 90°

Name ________________ Date ________ Class ________

LESSON 7-3

Problem Solving

Angles in Triangles

The American flag must be folded according to certain rules that result in the flag being folded into the shape of a triangle. The figure shows a frame designed to hold an American flag.

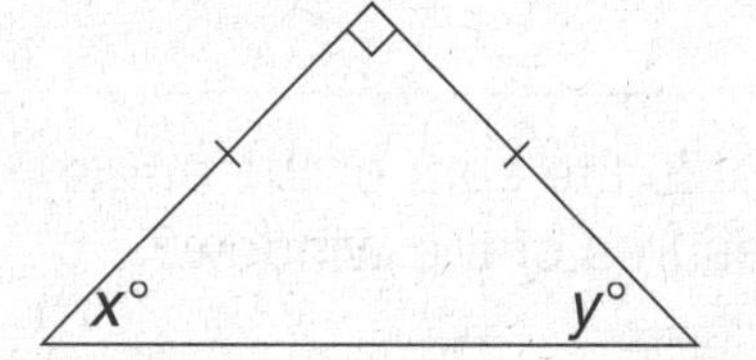

1. Is the triangle acute, right, or obtuse?

____

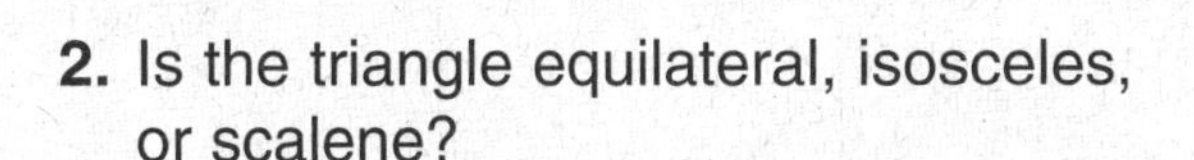

2. Is the triangle equilateral, isosceles, or scalene?

____

3. Find $x°$.

____

4. Find $y°$.

____

The figure shows a map of three streets. Choose the letter for the best answer.

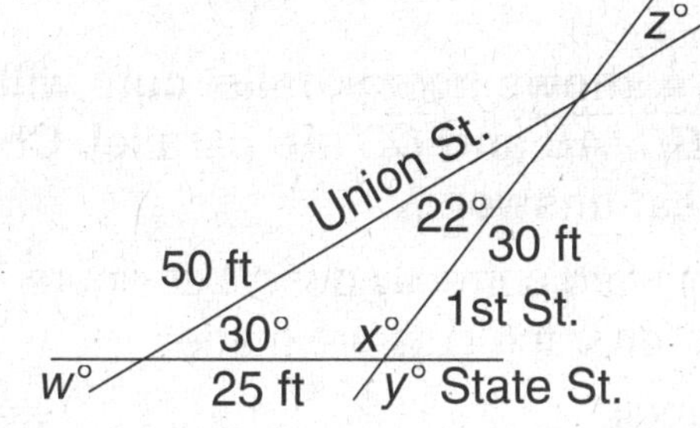

5. Find $x°$.

A 22°
B 128°
C 30°
D 68°

6. Find $w°$.

F 22°
G 128°
H 30°
J 52°

7. Find $y°$.

A 22°
B 30°
C 128°
D 143°

8. Find $z°$.

F 22°
G 30°
H 128°
J 143°

9. Which word best describes the triangle formed by the streets?

A acute
B right
C obtuse
D equilateral

10. Which word best describes the triangle formed by the streets?

F equilateral
G isosceles
H scalene
J acute

Name ______________________ Date ____________ Class ____________

LESSON 7-4

Problem Solving

Classifying Polygons

The figure shows how the glass for a window will be cut from a square piece. Cuts will be made along $\overline{CE}$, $\overline{FH}$, $\overline{IK}$, and $\overline{LB}$.

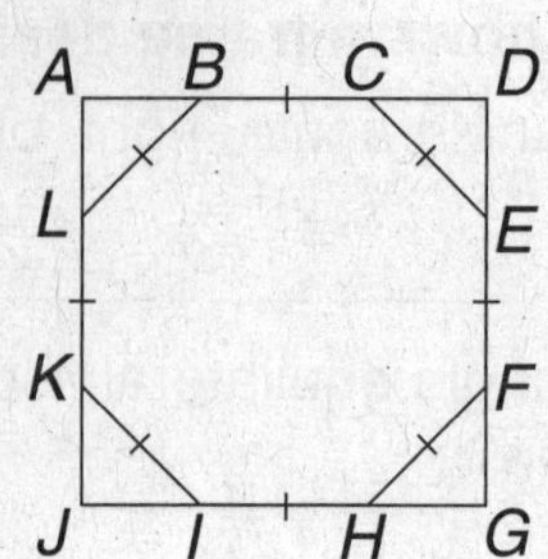

1. What shape is the window?

____

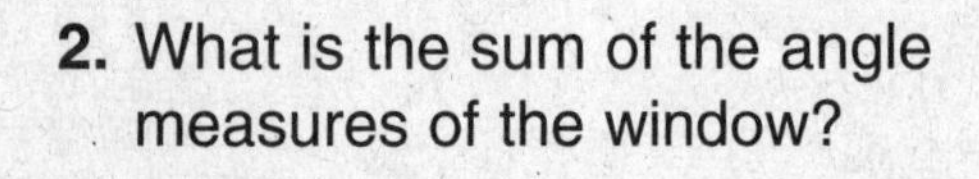

2. What is the sum of the angle measures of the window?

____

3. What is the measure of each angle of the window?

____

4. Based on the angles, what kind of triangle is $\triangle CDE$?

____

5. Based on the sides, what kind of triangle is $\triangle CDE$?

____

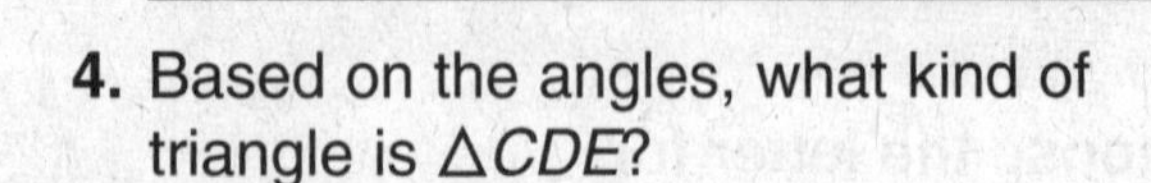

The figure shows how parallel cuts will be made along $\overline{AD}$ and $\overline{BC}$. $\overline{AB}$ and $\overline{CD}$ are parallel. Choose the letter for the best answer.

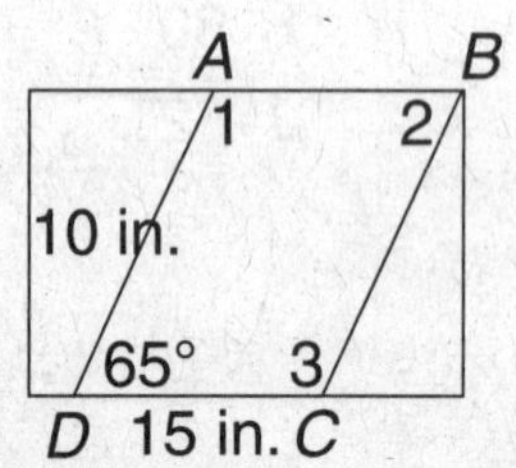

6. Which word correctly describes figure *ABCD* after the cuts are made?

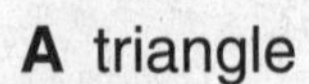

A triangle
B quadrilateral
C pentagon
D hexagon

7. Which word correctly describes figure *ABCD* after the cuts are made?

F parallelogram
G trapezoid
H rectangle
J rhombus

8. Find the measure of $\angle 1$.

A 45°
B 65°
C 90°
D 115°

9. Find the measure of $\angle 2$.

F 45° **H** 65°
G 90° **J** 115°

10. Find the measure of $\angle 3$.

A 45° **C** 65°
B 90° **D** 115°

Name ____________________ Date __________ Class __________

LESSON 7-5

Problem Solving

Coordinate Geometry

The Uniform Federal Accessibility Standards describes the standards for making buildings accessible for the handicapped. The standards say that the least possible slope should be used for a ramp and that the maximum slope of a ramp should be $\frac{1}{12}$.

1. What is the slope of the pictured ramp? Does the ramp meet the standard?

ramp
12 in.
12 ft

____

2. What is the slope of the pictured ramp? Does the ramp meet the standard?

ramp
12 in.
10 ft

____

Write the correct answer.

3. Find the slope of the roof.

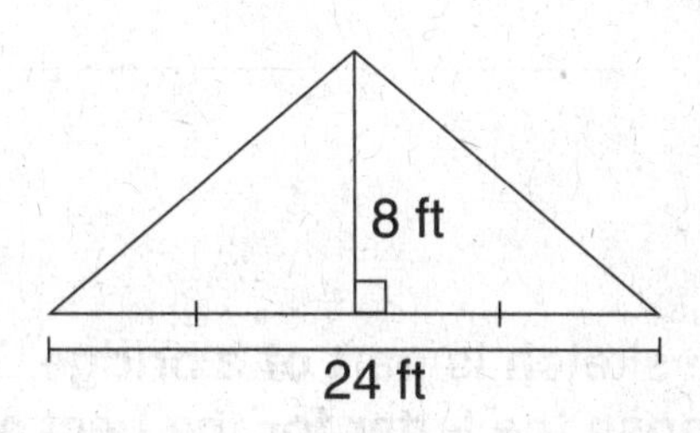

____

Choose the letter that represents the slope.

4. Many building codes require that a staircase be built with a maximum rise of 8.25 inches for a minimum tread width (run) of 9 inches.

   A $\frac{8}{9}$ **C** $\frac{9}{8.25}$

   B $\frac{11}{12}$ **D** $\frac{12}{11}$

5. Hills that have a rise of about 10 feet for every 17 feet horizontally are too steep for most cars.

   F $\frac{10}{17}$ **H** $\frac{17}{10}$

   G $\frac{2}{5}$ **J** $\frac{3}{5}$

6. At its steepest part, an intermediate ski run has a rise of about 4 feet for 10 feet horizontally.

   A $\frac{2}{5}$ **C** $\frac{5}{2}$

   B $\frac{4}{5}$ **D** $\frac{5}{4}$

7. Black diamond, or expert, ski slopes often have a rise of 10 feet for every 14 feet horizontally.

   F $\frac{7}{5}$ **H** $\frac{5}{7}$

   G $\frac{2}{7}$ **J** $\frac{7}{2}$

Name ______________________ Date ____________ Class ____________

LESSON **7-6**

Problem Solving

Congruence

Use the American patchwork quilt block design called Carnival to answer the questions. Triangle $AIH \cong$ Triangle AIB, Triangle $ACJ \cong$ Triangle AGJ, Triangle $GFJ \cong$ Triangle CDJ.

1. What is the measure of $\angle IAB$?

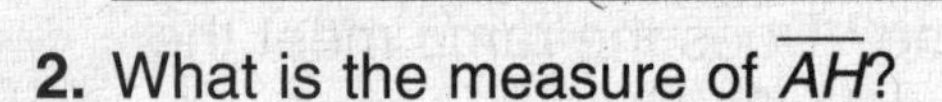

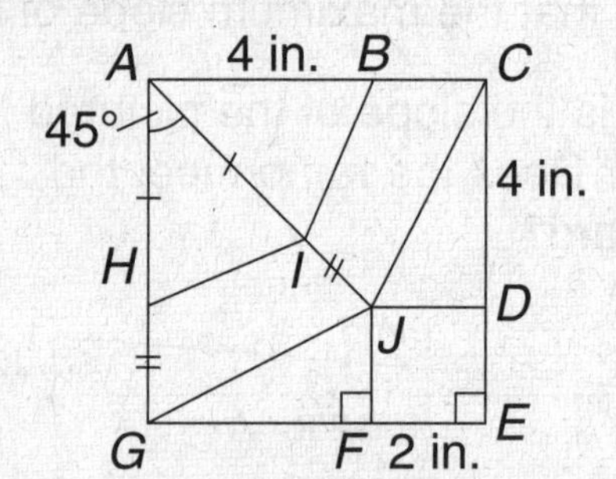

2. What is the measure of $\overline{AH}$?

____

3. What is the measure of $\overline{AG}$?

____

4. What is the measure of $\angle JDC$?

____

5. What is the measure of $\overline{FG}$?

____

The sketch is part of a bridge. Trapezoid $ABEF \cong$ Trapezoid $DEBC$. Choose the letter for the best answer.

6. What is the measure of $\overline{DE}$?

A 4 feet
B 8 feet
C 16 feet
D Cannot be determined

7. What is the measure of $\overline{FE}$?

F 4 feet
H 8 feet
G 16 feet
J 24 feet

8. What is the measure of $\angle FAB$?

A 45°
C 60°
B 90°
D 120°

9. What is the measure of $\angle ABE$?

F 45°
H 60°
G 90°
J 120°

10. What is the measure of $\angle EBC$?

A 45°
C 60°
B 90°
D 120°

11. What is the measure of $\angle BED$?

F 45°
H 60°
G 90°
J 120°

12. What is the measure of $\angle BCD$?

A 45°
C 60°
B 90°
D 120°

Name ______________________ Date ____________ Class ____________

LESSON 7-7

Problem Solving

Transformations

Parallelogram ABCD has vertices *A*(–3, 1), *B*(–2, 4), *C*(3, 4), and *D*(2, 1). Refer to the parallelogram to write the correct answer.

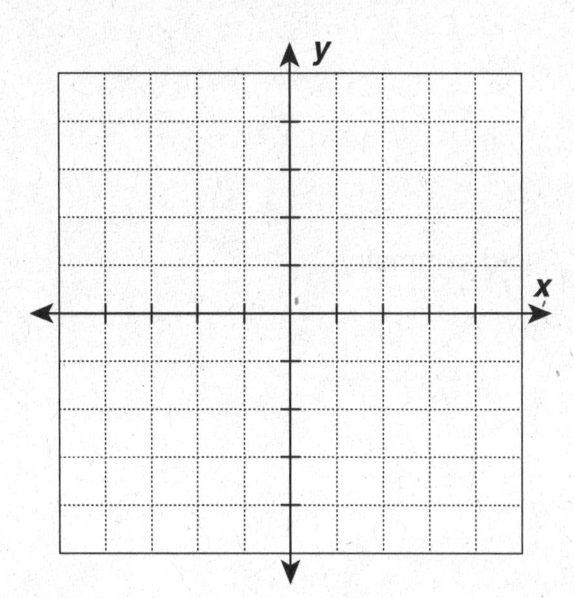

1. What are the coordinates of point A after a reflection across the x-axis?

   ____

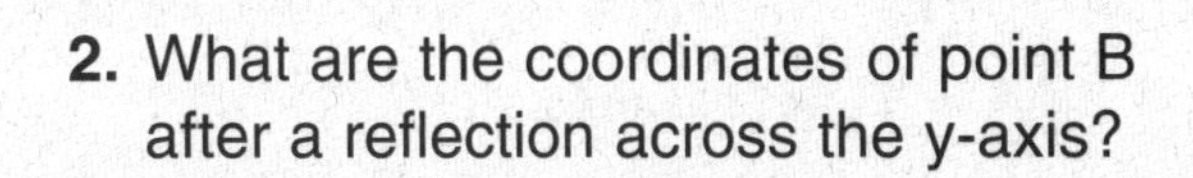

2. What are the coordinates of point B after a reflection across the y-axis?

   ____

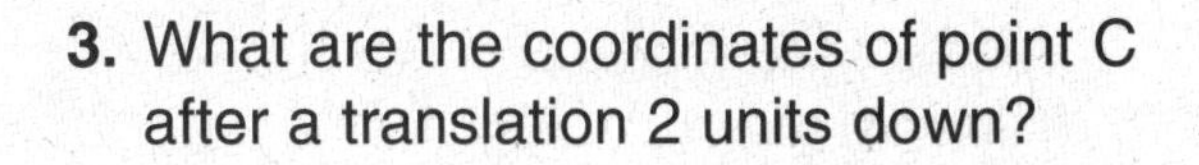

3. What are the coordinates of point C after a translation 2 units down?

   ____

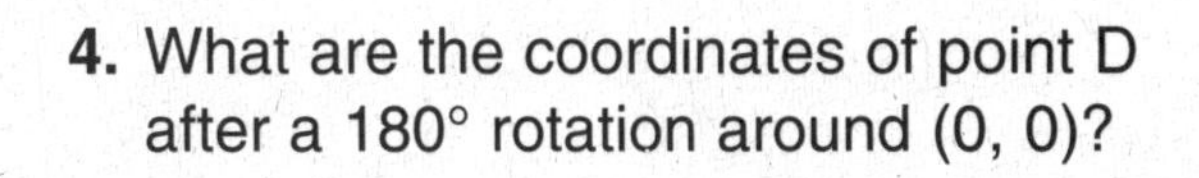

4. What are the coordinates of point D after a 180° rotation around (0, 0)?

   ____

Identify each as a translation, rotation, reflection or none of these.

5. 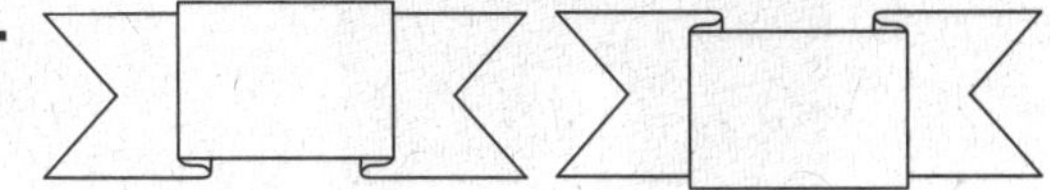

   A translation **C** rotation
   B reflection **D** none of these

6. 

   F translation **H** rotation
   G reflection **J** none of these

7. 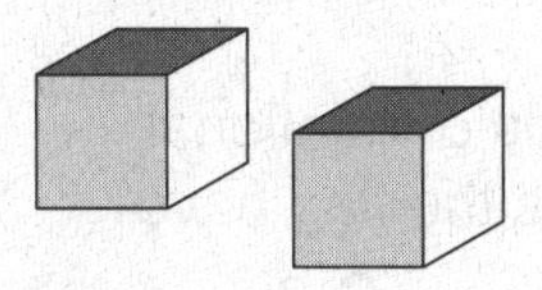

   A translation **C** rotation
   B reflection **D** none of these

8. **F** translation **H** rotation
   G reflection **J** none of these

Name ______________________________ Date ____________ Class ____________

LESSON 7-8

Problem Solving
Symmetry

Complete the figure. A dashed line is a line of symmetry and a point is a center of rotation.

1.

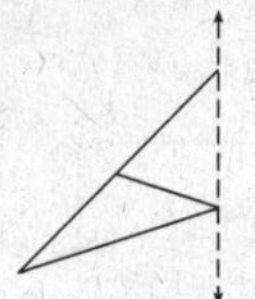

2.

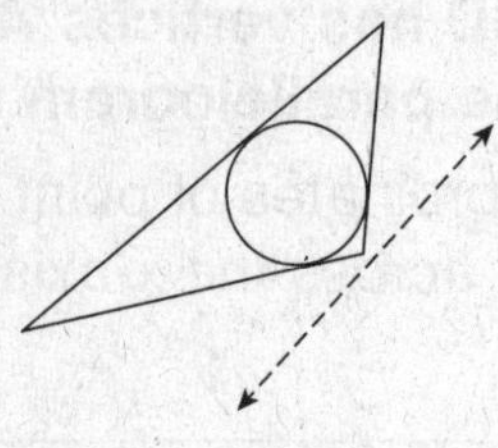

3. 2 fold symmetry

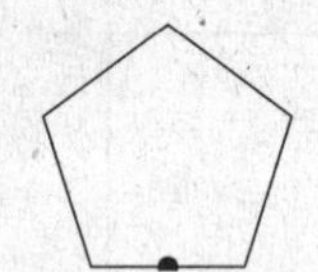

4. 4 fold symmetry

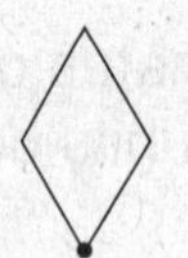

Use the flag of Switzerland to answer the questions.

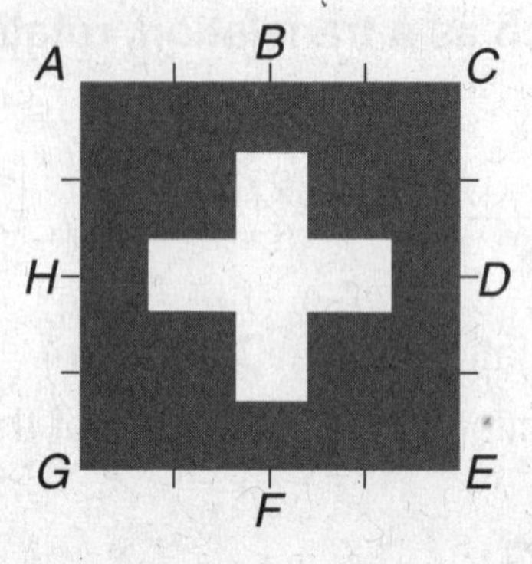

5. Which of the following would NOT be a line of symmetry?

A $\overline{HD}$ **C** $\overline{AE}$

B $\overline{BF}$ **D** $\overline{HB}$

6. How many lines of symmetry does the flag have?

F 2 **H** 4

G 6 **J** 8

7. How many folds of rotational symmetry does the flag have?

A 0 **C** 2

B 4 **D** 8

8. Which lists all lines of symmetry of the flag?

F $\overline{AE}$, $\overline{GC}$

G $\overline{HD}$, $\overline{BF}$

H $\overline{HD}$, $\overline{BF}$, $\overline{AE}$, $\overline{GC}$

J $\overline{HB}$, $\overline{DF}$, $\overline{AE}$, $\overline{GC}$

9. Which describes the center of rotation?

A intersection of $\overline{BF}$ and $\overline{HD}$

B intersection of $\overline{AE}$ and $\overline{HB}$

C *A*

D There is no center of rotation

Name ______________________ Date __________ Class __________

LESSON 7-9

Problem Solving

Tessellations

Create a tessellation using the given figure.

1. 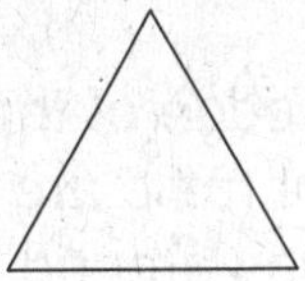

2. 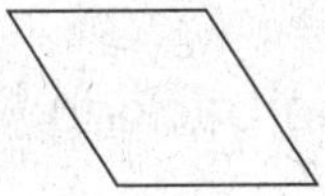

Choose the letter for the best answer.

3. Which figure will NOT make a tessellation?

A 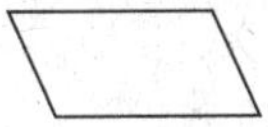

C 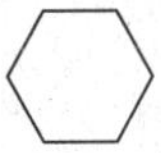

B 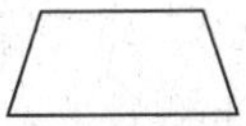

D 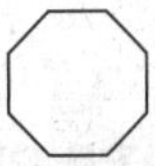

4. Which nonregular polygon can always be used to tile a floor?

F pentagon

G triangle

H octagon

J hexagon

5. For a combination of regular polygons to tessellate, the angles that meet at each vertex must add to what?

A 90°

B 180°

C 360°

D 720°

Name ______________________ Date ____________ Class ____________

LESSON 8-1

Problem Solving

Perimeter and Area of Rectangles and Parallelograms

Use the following for Exercises 1–2. A quilt for a twin bed is 68 in. by 90 in.

1. What is the area of the backing applied to the quilt?

____

2. A ruffle is sewn to the edge of the quilt. How many feet of ruffle are needed to go all the way around the edge of the quilt?

____

Use the following for Exercises 3–4. Jaime is building a rectangular dog run that is 12 ft by 8 ft.

3. If the run is cemented, how many square feet will be covered by cement?

____

4. How much fencing will be required to enclose the dog run?

____

Use the following for Exercises 5–6. Jackie is painting the walls in a room. Two walls are 12 ft by 8 ft, and two walls are 10 ft by 8 ft. Choose the letter for the best answer.

5. What is the area of the walls to be painted?

   A 352 ft^2 **C** 704 ft^2

   B 176 ft^2 **D** 400 ft^2

6. If a can of paint covers 300 square feet, how many cans of paint should Jackie buy?

   F 1 **H** 3

   G 2 **J** 4

Use the following for Exercises 7–8. One kind of pool cover is a tarp that stretches over the area of the pool and is tied down on the edge of the pool. The cover extends 6 inches beyond the edge of the pool. Choose the letter for the best answer.

7. A rectangular pool is 20 ft by 10 ft. What is the area of the tarp that will cover the pool?

   A 200 ft^2 **C** 60 ft^2

   B 231 ft^2 **D** 215.25 ft^2

8. If the tarp costs $2.50 per square foot, how much will the tarp cost?

   F $500.00 **H** $150.00

   G $538.13 **J** $577.50

Name ______________________ Date __________ Class __________

LESSON 8-2

Problem Solving

Perimeter and Area of Triangles and Trapezoids

Write the correct answer.

1. Find the area of the material required to cover the kite pictured below.

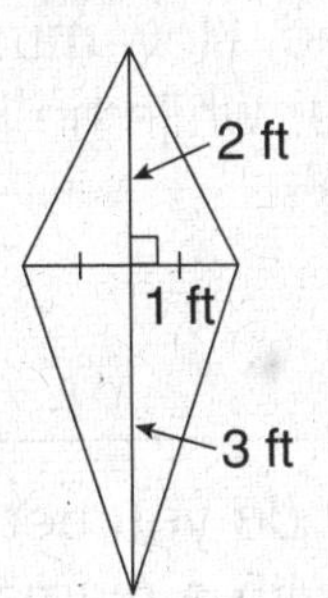

____

2. Find the area of the material required to cover the kite pictured below.

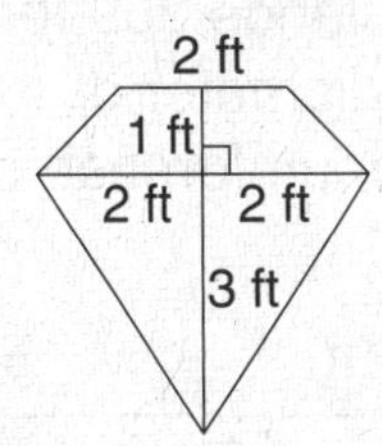

____

3. Find the approximate area of the state of Nevada.

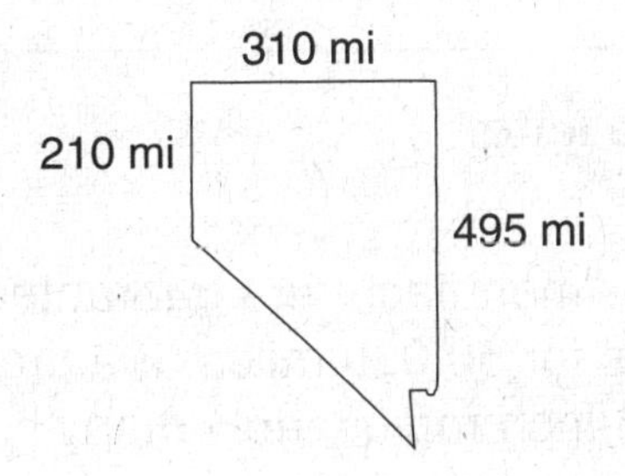

____

4. Find the area of the hexagonal gazebo floor.

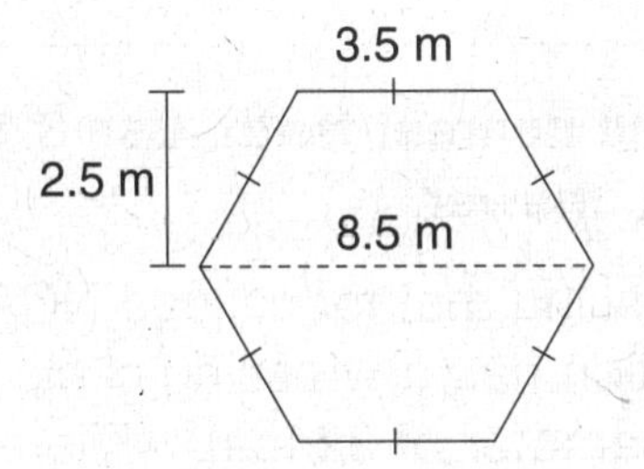

____

Choose the letter for the best answer.

5. Find the amount of flooring needed to cover the stage pictured below.

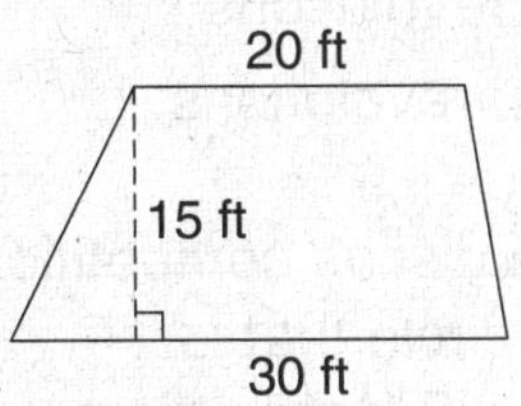

A 4500 ft^2

B 750 ft^2

C 525 ft^2

D 375 ft^2

6. Find the combined area of the congruent triangular gables.

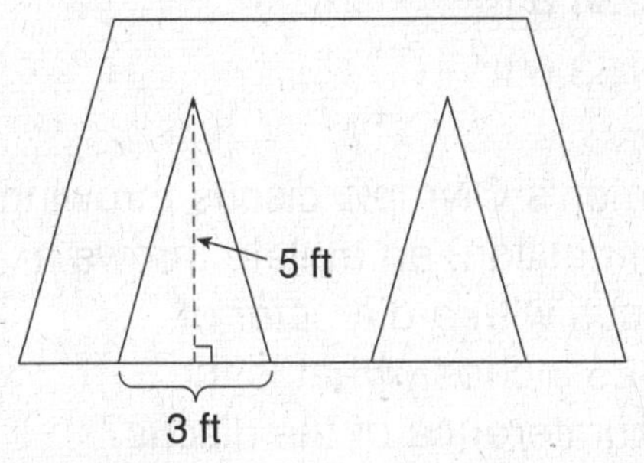

F 7.5 ft^2

G 15 ft^2

J 60 ft^2

H 30 ft^2

Name ______________________ Date ____________ Class ____________

LESSON 8-3

Problem Solving

Circles

Round to the nearest tenth. Use 3.14 for π. Write the correct answer.

1. The world's tallest Ferris wheel is in Osaka, Japan, and stands 369 feet tall. Its wheel has a diameter of 328 feet. Find the circumference of the Ferris wheel.

____

2. A dog is on a 15-foot chain that is anchored to the ground. How much area can the dog cover while he is on the chain?

____

3. A small pizza has a diameter of 10 inches, and a medium has a diameter of 12 inches. How much more pizza do you get with the medium pizza?

____

4. How much more crust do you get with a medium pizza with a diameter of 12 inches than a small pizza with a 10 inch diameter?

____

Round to the nearest tenth. Use 3.14 for π. Choose the letter for the best answer.

5. The wrestling mat for college NCAA competition has a wrestling circle with a diameter of 32 feet, while a high school mat has a diameter of 28 feet. How much more area is there in a college wrestling mat than a high school mat?

A 12.6 ft^2

B 188.4 ft^2

C 234.8 ft^2

D 753.6 ft^2

6. Many tire manufacturers guarantee their tires for 50,000 miles. If a tire has a 16-inch radius, how many revolutions of the tire are guaranteed? There are 63,360 inches in a mile. Round to the nearest revolution.

F 630.6 revolutions

G 3125 revolutions

H 31,528,662 revolutions

J 500,000,000 revolutions

7. In men's Olympic discus throwing competition, an athlete throws a discus with a diameter of 8,625 inches. What is the circumference of the discus?

A 13.5 in.

B 27.1 in.

C 58.4 in.

D 233.6 in.

8. An athlete in a discus competition throws from a circle that is approximately 8.2 feet in diameter. What is the area of the discus throwing circle?

F 52.8 ft^2

G 25.7 ft^2

H 12.9 ft^2

J 211.1 ft^2

Name ________________________ Date ____________ Class ____________

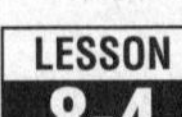

Problem Solving

Drawing Three-Dimensional Figures

Write the correct answer.

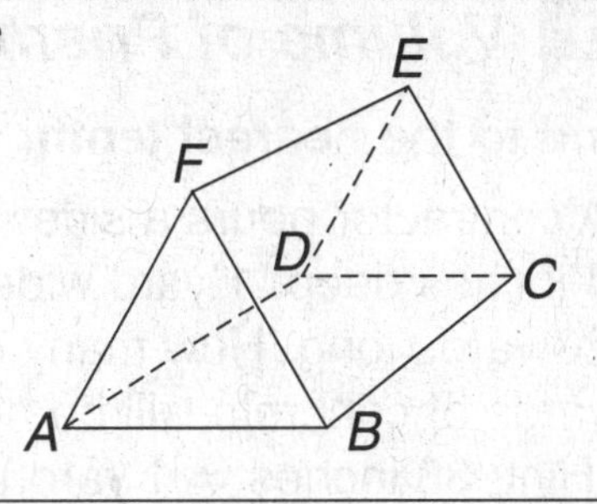

1. Mitch used a triangular prism in his science experiment. Name the vertices, edges, and faces of the triangular prism shown at the right.

____

____

____

2. Amber used cubes to make the model shown below of a sculpture she wants to make. Draw the front, top, and side views of the model.

Choose the letter of the best answer.

3. Which is the front view of the figure shown at the left? ____

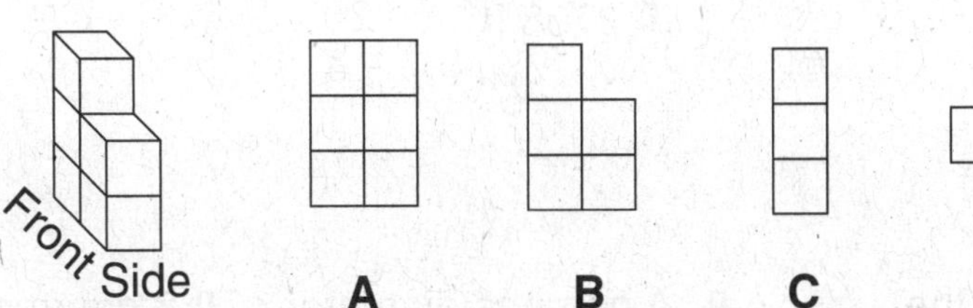

4. Which is **not** an edge of the figure shown at the right?

F $\overline{AB}$ **H** $\overline{EF}$

G $\overline{EH}$ **J** $\overline{BD}$

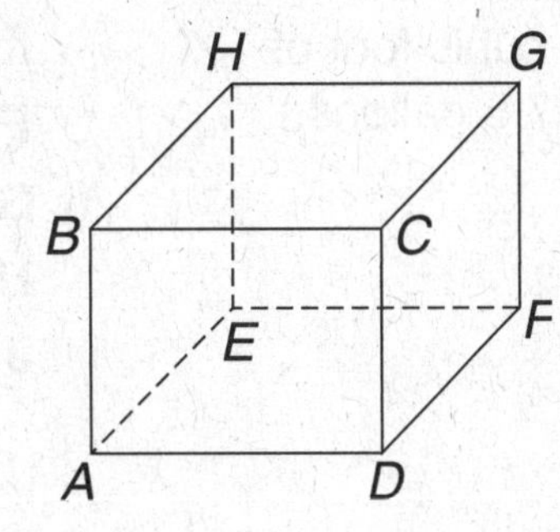

Name ______________________ Date ____________ Class ____________

LESSON 8-5

Problem Solving

Volume of Prisms and Cylinders

Round to the nearest tenth. Write the correct answer.

1. A contractor pours a sidewalk that is 4 inches deep, 1 yard wide, and 20 yards long. How many cubic yards of concrete will be needed? (Hint: 36 inches = 1 yard.)

____

2. A refrigerator has inside measurements of 50 cm by 118 cm by 44 cm. What is the capacity of the refrigerator?

____

A rectangular box is 2 inches high, 3.5 inches wide and 4 inches long. A cylindrical box is 3.5 inches high and has a diameter of 3.2 inches. Use 3.14 for π. Round to the nearest tenth.

3. Which box has a larger volume?

____

4. How much bigger is the larger box?

____

Use 3.14 for π. Choose the letter for the best answer.

5. A child's wading pool has a diameter of 5 feet and a height of 1 foot. How much water would it take to fill the pool? Round to the nearest gallon. (Hint: 1 cubic foot of water is approximately 7.5 gallons.)

   A 79 gallons
   B 589 gallons
   C 59 gallons
   D 147 gallons

6. How many cubic feet of air are in a room that is 15 feet long, 10 feet wide and 8 feet high?

   F 33 ft^3
   G 1200 ft^3
   H 1500 ft^3
   J 3768 ft^3

7. How many gallons of water will the water trough hold? Round to the nearest gallon. (Hint: 1 cubic foot of water is approximately 7.5 gallons.)

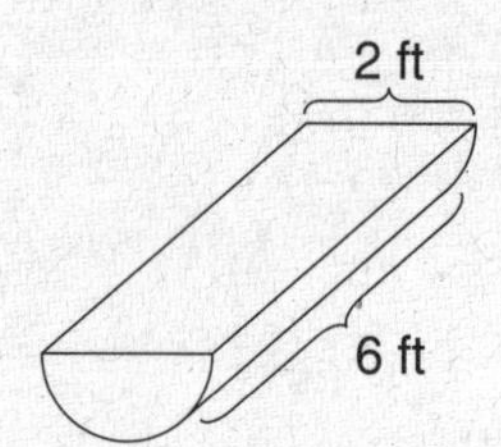

   A 19 gallons **C** 141 gallons
   B 71 gallons **D** 565 gallons

8. A can has diameter of 9.8 cm and is 13.2 cm tall. What is the capacity of the can? Round to the nearest tenth.

   F 203.1 cm^3
   G 995.2 cm^3
   H 3980.7 cm^3
   J 959.2 cm^3

Name ______________________ Date __________ Class __________

LESSON 8-6

Problem Solving

Volume of Pyramids and Cones

Round to the nearest tenth. Use 3.14 for π. Write the correct answer.

1. The Feathered Serpent Pyramid in Teotihuacan, Mexico is the third largest in the city. Its base is a square that measures 65 m on each side. The pyramid is 19.4 m high. What is the volume of the Feathered Serpent Pyramid?

____

2. The Sun Pyramid in Teotihuacan, Mexico, is larger than the Feathered Serpent Pyramid. The sides of the square base and the height are each about 3.3 times larger than the Feathered Serpent Pyramid. How many times larger is the volume of the Sun Pyramid than the Feathered Serpent Pyramid?

____

3. An oil funnel is in the shape of a cone. It has a diameter of 4 inches and a height of 6 inches. If the end of the funnel is plugged, how much oil can the funnel hold before it overflows?

____

4. One quart of oil has a volume of approximately 57.6 in^3. Does the oil funnel in exercise 3 hold more or less than 1 quart of oil?

____

Round to the nearest tenth. Use 3.14 for π. Choose the letter for the best answer.

5. An ice cream cone has a diameter of 4.2 cm and a height of 11.5 cm. What is the volume of the cone?

A 18.7 cm^3

B 25.3 cm^3

C 53.1 cm^3

D 212.3 cm^3

6. When decorating a cake, the frosting is put into a cone shaped bag and then squeezed out a hole at the tip of the cone. How much frosting is in a bag that has a radius of 1.5 inches and a height of 8.5 inches?

F 5.0 in^3

G 13.3 in^3

H 15.2 in^3

J 20.0 in^3

7. What is the volume of the hourglass at the right?

A 13.1 in^3

B 26.2 in^3

C 52.3 in^3

D 102.8 in^3

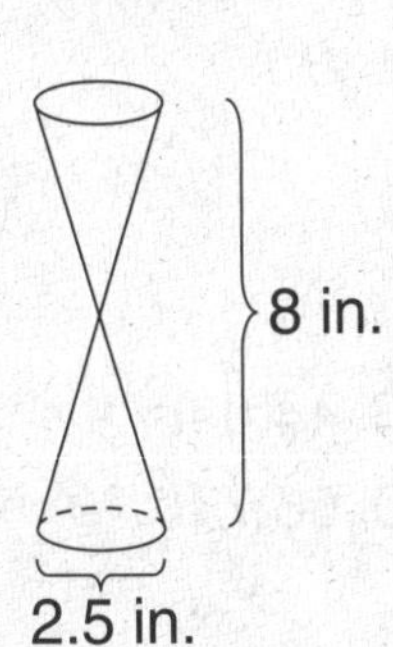

Name ______________________ Date __________ Class __________

LESSON **8-7**

Problem Solving
Surface Area of Prisms and Cylinders

An important factor in designing packaging for a product is the amount of material required to make the package. Consider the three figures described in the table below. Use 3.14 for π. Round to the nearest tenth. Write the correct answer.

1. Find the surface area of each package given in the table.

2. Which package has the lowest materials cost? Assume all of the packages are made from the same material.

____

Package	Dimensions	Volume	Surface Area
Prism	Base: 2" × 16" Height = 2"	64 in^3	
Prism	Base: 4" × 4" Height = 4"	64 in^3	
Cylinder	Radius = 2" Height = 5.1"	64.06 in^3	

Use 3.14 for π. Round to the nearest hundredth.

3. How much cardboard material is required to make a cylindrical oatmeal container that has a diameter of 12.5 cm and a height of 24 cm, assuming there is no overlap? The container will have a plastic lid.

____

4. What is the surface area of a rectangular prism that is 5 feet by 6 feet by 10 feet?

____

Use 3.14 for π. Round to the nearest tenth. Choose the letter for the best answer.

5. How much metal is required to make the trough pictured below?

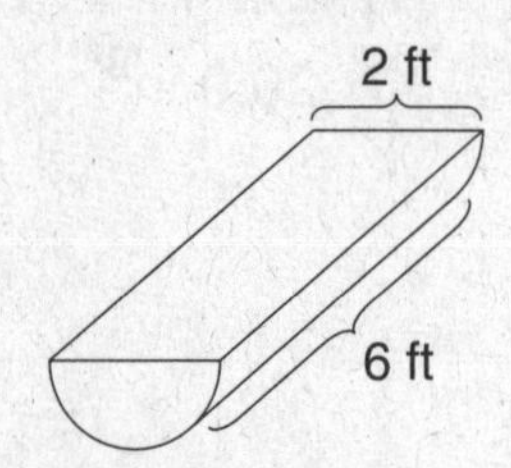

A 22.0 ft^2
B 34.0 ft^2
C 44.0 ft^2
D 56.7 ft^2

6. A can of vegetables has a diameter of 9.8 cm and is 13.2 cm tall. How much paper is required to make the label, assuming there is no overlap? Round to the nearest tenth.

F 203.1 cm^2
G 406.2 cm^2
H 557.0 cm^2
J 812.4 cm^2

Name _______________ Date _______ Class _______

LESSON 8-8

Problem Solving

Surface Area of Pyramids and Cones

Round to the nearest tenth. Use 3.14 for π. Write the correct answer.

1. The Feathered Serpent Pyramid in Teotihuacan, Mexico, is the third largest in the city. Its base is a square that measures 65 m on each side. The pyramid is 19.4 m high and has a slant height of 37.8 m. The lateral faces of the pyramid are decorated with paintings. What is the surface area of the painted faces?

____

2. The Sun Pyramid in Teotihuacan, Mexico, is larger than the Feathered Serpent Pyramid. The sides of the square base and the slant height are each about 3.3 times larger than the Feathered Serpent Pyramid. How many times larger is the surface area of the lateral faces of the Sun Pyramid than the Feathered Serpent Pyramid?

____

3. An oil funnel is in the shape of a cone. It has a diameter of 4 inches and a slant height of 6 inches. How much material does it take to make a funnel with these dimensions?

____

4. If the diameter of the funnel in Exercise 6 is doubled, by how much does it increase the surface area of the funnel?

____

Round to the nearest tenth. Use 3.14 for π. Choose the letter for the best answer.

5. An ice cream cone has a diameter of 4.2 cm and a slant height of 11.5 cm. What is the surface area of the ice cream cone?

   A 4.7 cm^2 **C** 75.83 cm^2
   B 19.9 cm^2 **D** 159.2 cm^2

6. A marker has a conical tip. The diameter of the tip is 1 cm and the slant height is 0.7 cm. What is the area of the writing surface of the marker tip?

   F 1.1 cm^2 **H** 2.2 cm^2
   G 1.9 cm^2 **J** 5.3 cm^2

7. A skylight is shaped like a square pyramid. Each panel has a 4 m base. The slant height is 2 m, and the base is open. The installation cost is $5.25 per square meter. What is the cost to install 4 skylights?

   A $64 **C** $218
   B $159 **D** $336

8. A paper drinking cup shaped like a cone has a 10 cm slant height and an 8 cm diameter. What is the surface area of the cone?

   F 88.9 cm^2 **H** 251.2 cm^2
   G 125.6 cm^2 **J** 301.2 cm^2

Name ______________________ Date __________ Class __________

LESSON 8-9

Problem Solving

Spheres

Early golf balls were smooth spheres. Later it was discovered that golf balls flew better when they were dimpled. On January 1, 1932, the United States Golf Association set standards for the weight and size of a golf ball. The minimum diameter of a regulation golf ball is 1.680 inches. Use 3.14 for π. Round to the nearest hundredth.

1. Find the volume of a smooth golf ball with the minimum diameter allowed by the United States Golf Association.

   ____

2. Find the surface area of a smooth golf ball with the minimum diameter allowed by the United States Golf Association.

   ____

3. Would the dimples on a golf ball increase or decrease the volume of the ball?

   ____

4. Would the dimples on a golf ball increase or decrease the surface area of the ball?

   ____

Use 3.14 for π. Use the following information for Exercises 5–6. A track and field expert recommends changes to the size of a shot put. One recommendation is that a shot put should have a diameter between 90 and 110 mm. Choose the letter for the best answer.

5. Find the surface area of a shot put with a diameter of 90 mm.

   A 25,434 mm^2
   B 101,736 mm^2
   C 381,520 mm^2
   D 3,052,080 mm^2

6. Find the surface area of a shot put with diameter 110 m.

   F 9,499 mm^2
   G 22,834 mm^2
   H 37,994 mm^2
   J 151,976 mm^2

7. Find the volume of the earth if the average diameter of the earth is 7926 miles.

   A 2.0×10^8 mi^3
   B 2.6×10^{11} mi^3
   C 7.9×10^8 mi^3
   D 2.1×10^{12} mi^3

8. An ice cream cone has a diameter of 4.2 cm and a height of 11.5 cm. One spherical scoop of ice cream is put on the cone that has a diameter of 5.6 cm. If the ice cream were to melt in the cone, by how much of it would overflow the cone? Round to the nearest tenth.

   F 0 cm^3 **H** 38.8 cm^3
   G 12.3 cm^3 **J** 54.3 cm^3

Name ______________________ Date __________ Class __________

LESSON **8-10**

Problem Solving

Scaling Three-Dimensional Figures

Round to the nearest hundredth. Write the correct answer.

1. The smallest regulation golf ball has a volume of 2.48 cubic inches. If the diameter of the ball were increased by 10%, or a factor of 1.1, what will the volume of the golf ball be?

____

2. The smallest regulation golf ball has a surface area of 8.86 square inches. If the diameter of the ball were increased by 10%, what will the surface area of the golf ball be?

____

3. The Feathered Serpent Pyramid in Teotihuacan, Mexico, is the third largest in the city. The dimensions of the Sun Pyramid in Teotihuacan, Mexico, are about 3.3 times larger than the Feathered Serpent Pyramid. How many times larger is the volume of the Sun Pyramid than the Feathered Serpent Pyramid?

____

4. The faces of the Feathered Serpent Pyramid and the Sun Pyramid described in Exercise 3 have ancient paintings on them. How many times larger is the surface are of the faces of the Sun Pyramid than the faces of the Feathered Serpent Pyramid?

____

Choose the letter for the best answer.

5. John is designing a shipping container that boxes will be packed into. The container he designed will hold 24 boxes. If he doubles the sides of his container, how many times more boxes will the shipping container hold?

A 2 **C** 8

B 4 **D** 192

6. If John doubles the sides of his container from exercise 5, how many times more material will be required to make the container?

F 2 **H** 8

G 4 **J** 192

7. A child's sandbox is shaped like a rectangular prism and holds 2 cubic feet of sand. The dimensions of the next size sandbox are double the smaller sandbox. How much sand will the larger sandbox hold?

A 4 ft^3 **C** 16 ft^3

B 8 ft^3 **D** 32 ft^3

8. Maria used two boxes of sugar cubes to create a solid building for a class project. She decides that the building is too small and she will rebuild it 3 times larger. How many more boxes of sugar cubes will she need?

F 4 **H** 27

G 25 **J** 52

Name ______________________ Date ____________ Class ____________

LESSON 9-1

Problem Solving

Samples and Surveys

Identify the sampling method used.

1. Every twentieth student on a list is chosen to participate in a poll.

____

2. Seat numbers are drawn from a hat to identify passengers on an airplane that will be surveyed.

____

Give a reason why the sample could be biased.

3. A company wants to find out how its customers rate their products. They ask people who visit the company's Web Site to rate their products.

____

____

____

4. A teacher polls all of the students who are in detention on Friday about their opinions on the amount of homework students should have each night.

____

____

____

A car dealership wants to know how people who have visited the dealership feel about the dealership and the sales people. They survey every 5th person who buys a car. Choose the letter for the best answer.

5. Identify the population.
   - **A** People who visit the dealership
   - **B** People who buy a car from the dealership
   - **C** People in the local area
   - **D** The salesmen at the dealership

6. Identify the sample.
   - **F** Every person who visits the dealership
   - **G** People who buy a car
   - **H** Every 5th buyer
   - **J** People in the local area

7. Identify the possible bias.
   - **A** Not all people will visit the dealership.
   - **B** Did not survey everyone who buys a car.
   - **C** Not including those who visited but did not buy.
   - **D** There is no bias.

8. Identify the sampling method used.
   - **F** Random
   - **G** Systematic
   - **H** Stratified
   - **J** None of these

Name ______________________ Date __________ Class __________

LESSON 9-2

Problem Solving

Organizing Data

A consumer survey gathered the following data about what teens do while on online.

1. Make a stem-and-leaf plot of the data.

Teens' Activities Online	
Activity	Percent
E-mail	95
Use search engines	86
Instant Messaging	82
Visit music sites	73
Enter contests	73

The stem-and-leaf plot that shows the total number of medals won by different countries in the 2000 Summer Olympics. Choose the letter for the best answer.

2000 Olympic Medals

Stem	Leaves
2	3 5 6 8 8 9
3	4 8
4	
5	7 8 9
6	
7	
8	8
9	7

2. List all the data values in the stem-and-leaf plot.

A 2, 4, 5, 6, 7, 8, 9

B 23, 25, 26, 28, 28, 29, 34, 38, 40, 57, 58, 59, 60, 70, 88, 97

C 23, 25, 26, 28, 29, 34, 38, 57, 58, 59, 88, 97

D 23, 25, 26, 28, 28, 29, 34, 38, 57, 58, 59, 88, 97

3. What is the least number of medals won by a country represented in the stem-and-leaf plot?

F 3

G 4

H 23

J 97

4. What is the greatest number of medals won by a country represented in the stem-and-leaf plot?

A 9

B 70

C 79

D 97

Name ______________________ Date __________ Class __________

LESSON 9-3

Problem Solving

Measures of Central Tendency

Use the data to find each answer.

World's Busiest Airports

Airport	Total Passengers (in millions)
Atlanta, Hartsfield	80.2
Chicago, O'Hare	72.1
Los Angeles	68.5
London, Heathrow	64.6
Dallas/Ft. Worth	60.7

1. Find the average number of passengers in the world's five busiest airports.

____

2. Find the median number of passengers in the world's five busiest airports.

____

3. Find the mode of the airport data.

____

4. Find the range of the airport data.

____

Choose the letter for the best answer.

World Motor Vehicle Production (in thousands) 1998–1999

Country	1998	1999
United States	12,047	13,063
Canada	2,568	3,026
Europe	16,332	16,546
Japan	10,050	9904

5. What was the mean production of motor vehicles in 1998?

A 8,651,500 vehicles
B 10,249,250 vehicles
C 11,264,250 vehicles
D 12,000,000 vehicles

6. What was the range of production in 1999?

F 9,800,000 vehicles
G 11,480,000 vehicles
H 12,520,000 vehicles
J 13,520,000 vehicles

7. What was the median number of vehicles produced in 1999?

A 3,026,000 vehicles
B 3,069,000 vehicles
C 11,483,500 vehicles
D 13,063,000 vehicles

8. Which value is largest?

F Mean of 1998 data
G Mean of 1999 data
H Median of 1998 data
J Median of 1999 data

Name ______________________ Date __________ Class __________

LESSON 9-4

Problem Solving

Variability

Write the correct answer.

1. Find the median of the data.

____

2. Find the first and third quartiles of the data.

____

3. Make a box-and-whisker plot of the data.

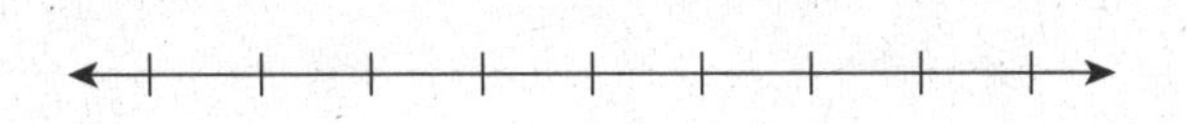

Super Bowl Point Differences

Year	Point Difference
2001	27
2000	7
1999	15
1998	7
1997	14
1996	10
1995	23
1994	17
1993	35
1992	13

The box-and-whisker plots compare the highest recorded Fahrenheit temperatures on the seven continents with the lowest recorded temperatures. Choose the letter for the best answer.

4. Which statement is true?

A The median of the high temperatures is less than the median of the low temperatures.

B The range of low temperatures is greater than the range of high temperatures.

C The range of the middle half of the data is greater for the high temperatures.

D The median of the high temperatures is 49°F.

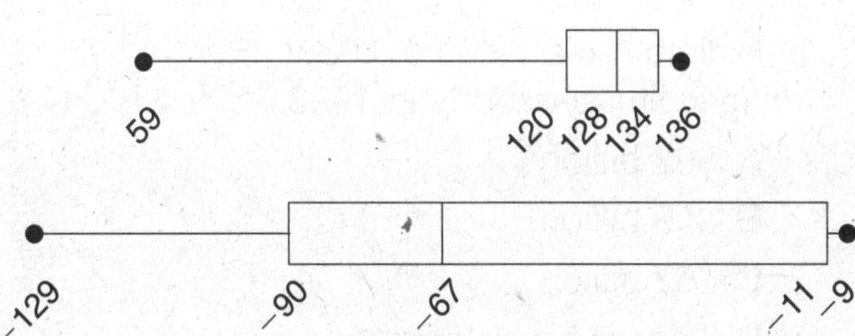

5. What is the median of the high temperatures?

F 128°F **H** –67°F

G 120°F **J** –90°F

6. What is the range of the low temperatures?

A 77°F **C** 120°F

B 79°F **D** 129°F

Name ______________________ Date __________ Class __________

LESSON **9-5**

Problem Solving

Displaying Data

Make the indicated graph.

1. Make a double-bar graph of the homework data.

Hours of Daily Homework	1	2	3	4	5
Boys	12	5	2	1	0
Girls	4	6	5	3	2

Hours of Daily Homework

Frequency

Hours

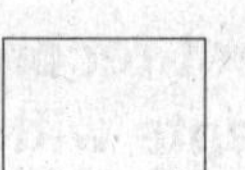

2. The annual hourly delay per driver in the 20 U.S. cities with the most traffic are as follows: 56, 42, 53, 46, 34, 37, 42, 34, 53, 21, 45, 50, 34, 42, 41, 38, 42, 34, 38, 31. Make a histogram with intervals of 5 hours.

For 3–5, refer to the double-line graph. Circle the letter of the correct answer.

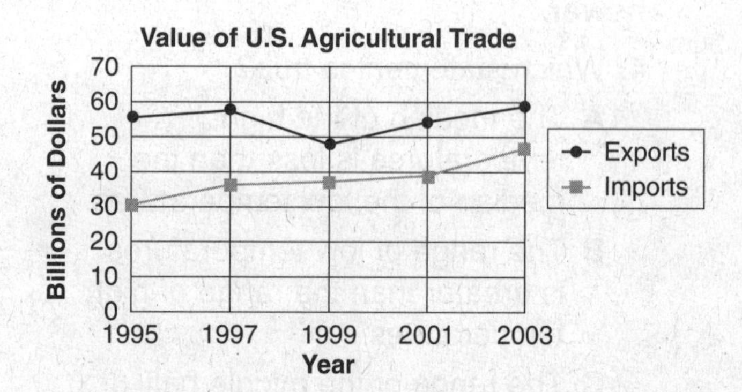

3. Estimate the value of U.S. agricultural exports in 1998.
   A $62 billion
   B $59 billion
   C $52 billion
   D Cannot be determined

4. Estimate the value of U.S. agricultural imports in 2000.
   F $39 billion
   G $31 billion
   H $29 billion
   J $21 billion

5. Estimate the difference between agricultural exports and imports in 1995.
   A $16 billion
   B $21 billion
   C $26 billion
   D Cannot be determined

Name ______________________ Date __________ Class __________

LESSON 9-6

Problem Solving

Misleading Graphs and Statistics

Explain why each statistics is misleading.

1. A poll taken at a college says that 38% of students like pizza the best, 32% like hamburgers the best, and 30% like spaghetti the best. They conclude that most of the students at the college like pizza the best.

____

____

2. The National Safety Council of Ireland found that young men were responsible in 57% of automobile accidents they were involved in. The NSC Web site made this claim: "Young men are responsible for over half of all road accidents."

____

____

3. Explain why the Centers for Disease Control (CDC) has been highly criticized for the graph below.

____

____

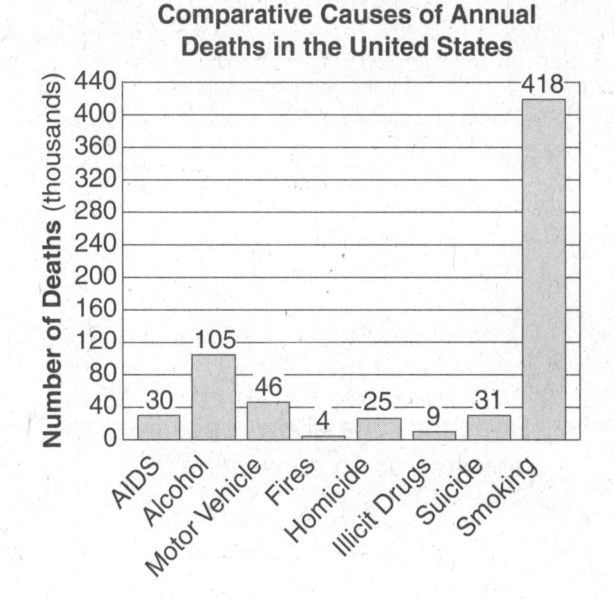

Choose the letter for the best answer.

4. Which statement is a misleading statistic for the data in the table?

Student	Test Grade
A	85%
B	92%
C	88%
D	10%
E	80%

A The median score was 85%.
B Most students scored an 80% or above.
C The average test score was 71%.
D The range of test scores was 82.

5. A sno-cone store claims, "Our sales have tripled!" Sno-cone sales from March to May were 50 and sales from June to August were 150. Why is this misleading?

A Sample size is too small.
B During the summer, sales should be higher.
C Should use the median not mean.
D The statement isn't misleading.

Name _______________________ Date __________ Class __________

LESSON 9-7

Problem Solving
Scatter Plots

Use the data given at the right.

1. Make a scatter plot of the data.

Percent of Americans Who Have Completed High School

Year	Percent
1910	13.5
1920	16.4
1930	19.1
1940	24.5
1950	34.3
1960	41.1
1970	55.2
1980	68.6
1990	77.6
1999	83.4

2. Does the data show a positive, negative or no correlation?

____

3. Use the scatter plot to predict the percent of Americans who will complete high school in 2010.

____

Choose the letter for the best answer.

4. Which data sets have a positive correlation?
   - **A** The length of the lines at amusement park rides and the number of rides you can ride in a day
   - **B** The temperature on a summer day and the number of visitors at a swimming pool
   - **C** The square miles of a state and the population of the state in the 2000 census
   - **D** The length of time spent studying and doing homework and the length of time spent doing other activities

5. Which data sets have a negative correlation?
   - **F** The number of visitors at an amusement park and the length of the lines for the rides
   - **G** The amount of speed over the speed limit when you get a speeding ticket and the amount of the fine for speeding
   - **H** The temperature and the number of people wearing coats
   - **J** The distance you live from school and the amount of time it takes to get to school

Name ______________________ Date __________ Class __________

LESSON 9-8

Problem Solving

Choosing the Best Representation of Data

Write what kind of graph would be best to display the described data.

1. Numbers of times that members of track team ran a mile in the following intervals: 4 min 31 s to 4 min 40 s, 4 min 41 s to 4 min 50 s, 4 min 51 s to 5 min, 5 min 1 s to 5 min 10 s

____

2. Distribution and range of students' scores on a history exam

____

3. Relationship between the amounts of time a student spent on her math homework and the numbers of homework problems she solved

____

4. Total numbers of victories of eight teams in an intramural volleyball league

____

5. Part of calories in a meal that come from protein

____

6. Numbers of books that a student reads each month over a year

____

Choose the letter for the best answer.

7. A bar graph is a good way to display

A data that changes over time.

B parts of a whole.

C distribution of data.

D comparison of different groups of data.

8. A circle graph is a good way to display

F range and distribution of data.

G distribution of data.

H parts of a whole.

J changes in data over time.

9. A scatter plot is a good way to display

A comparison of different groups of data.

B distribution and range of data.

C the relationship between two sets of data.

D parts of a whole.

10. A box-and-whisker plot is a good way to display

F range and distribution of data.

G the relationship between two sets of data.

H data that changes over time.

J parts of a whole.

Name ______________________ Date ____________ Class ____________

LESSON 10-1

Problem Solving

Probability

Write the correct answer.

1. To get people to buy more of their product, a company advertises that in selected boxes of their popsicles is a super hero trading card. There is a $\frac{1}{4}$ chance of getting a trading card in a box. What is the probability that there will not be a trading card in the box of popsicles that you buy?

   ____

2. The probability of winning a lucky wheel television game show in which 6 preselected numbers are spun on a wheel numbered 1–49 is $\frac{1}{13{,}983{,}816}$ or 0.000007151%. What is the probability that you will not win the game show?

   ____

Based on world statistics, the probability of identical twins is 0.004, while the probability of fraternal twins is 0.023.

3. What is the probability that a person chosen at random from the world will be a twin?

   ____

4. What is the probability that a person chosen at random from the world will not be a twin?

   ____

Use the table below that shows the probability of multiple births by country. Choose the letter for the best answer.

Probability of Multiple Births

Country	Probability
Canada	0.012
Japan	0.008
United Kingdom	0.014
United States	0.029
Sweden	0.014
Switzerland	0.013

5. In which country is it most likely to have multiple births?
   A Japan **C** Sweden
   B United States **D** Switzerland

6. In which country is it least likely to have multiple births?
   F Japan **H** Sweden
   G United States **J** Switzerland

7. In which two countries are multiple births equally likely?
   A United Kingdom, Canada
   B Canada, Switzerland
   C Sweden, United Kingdom
   D Japan, United States

Name ______________________ Date __________ Class __________

LESSON 10-2

Problem Solving

Experimental Probability

Use the table below. Round to the nearest percent. Write the correct answer.

Average Number of Days of Sunshine Per Year for Selected Cities

City	Number of Days
Buffalo, NY	175
Fort Wayne, IN	215
Miami, FL	256
Raleigh, NC	212
Richmond, VA	230

1. Estimate the probability of sunshine in Buffalo, NY.

____

2. Estimate the probability of sunshine in Fort Wayne, IN.

____

3. Estimate the probability of sunshine in Miami, FL.

____

4. Estimate the probability that it will not be sunny in Raleigh, NC.

____

5. Estimate the probability that it will not be sunny in Miami, FL.

____

6. Estimate the probability of sunshine in Richmond, VA.

____

Use the table below that shows the number of deaths and injuries caused by lightning strikes. Choose the letter for the best answer.

States with Most Lightning Deaths

State	Average deaths per year	Average injuries per year	Population
Florida	9.6	32.7	15,982,378
North Carolina	4.6	12.9	8,049,313
Texas	4.6	9.3	20,851,820
New York	3.6	12.5	18,976,457
Tennessee	3.4	9.7	5,689,283

7. Estimate the probability of being injured by a lightning strike in New York.

A 0.0000007% **C** 0.00007%

B 0.0000002% **D** 0.000002%

8. Estimate the probability of being killed by lightning in North Carolina.

F 0.0000006% **H** 0.00002%

G 0.00006% **J** 0.000002%

9. Estimate the probability of being struck by lightning in Florida.

A 0.00006%

B 0.00026%

C 0.0000026%

D 0.0006%

10. In which two states do you have the highest probability of being struck by lightning?

F Florida, North Carolina

G Florida, Tennessee

H Texas, New York

J North Carolina, Tennessee

Name ______________________ Date __________ Class __________

LESSON 10-3

Problem Solving

Use a Simulation

Use the table of random numbers below. Use at least 10 trials to simulate each situation. Write the correct answer.

87244	11632	85815	61766
19579	28186	18533	24633
74581	65633	54238	32848
87549	85976	13355	46498
53736	21616	86318	77291
24794	31119	48193	44869
86585	27919	65264	93557
94425	13325	16635	25840
18394	73266	67899	38783
94228	23426	76679	41256

1. Of people 18–24 years of age, 49% do volunteer work. If 10 people ages 18–24 were chosen at random, estimate the probability that at least 4 of them do volunteer work.

____

2. In the 2000 Presidential election, 56% of the population of North Carolina voted for George W. Bush. If 10 people were chosen at random from North Carolina, estimate the probability that at least 8 of them voted for Bush.

____

3. Forty percent of households with televisions watched the 2001 Super Bowl game. If 10 households with televisions are chosen at random, estimate the probability that at least 3 watched the 2001 Super Bowl.

____

Use the table above and at least 10 trials to simulate each situation. Choose the letter for the best estimate.

4. As of August 2000, 42% of U.S. households had Internet access. If 10 households are chosen at random, estimate the probability that at least 5 of them will have Internet access.

A 0% **C** 60%

B 30% **D** 90%

5. On average, there is rain 20% of the days in April in Orlando, FL. Estimate the probability that it will rain at least once during your 7-day vacation in Orlando in April.

F 20% **H** 70%

G 50% **J** 40%

6. Kareem Abdul-Jabaar is the NBA lifetime leader in field goals. During his career, he made 56% of the field goals he attempted. In a given game, estimate the probability that he would make at least 6 out of 10 field goals.

A 40% **C** 80%

B 60% **D** 100%

7. At the University of Virginia 39% of the applicants are accepted. If 10 applicants to the University of Virginia are chosen at random, estimate the probability that at least 4 of them are accepted to the University of Virginia.

F 10% **H** 80%

G 40% **J** 70%

Name ______________________ Date ___________ Class ___________

LESSON 10-4

Problem Solving

Theoretical Probability

A company that sells frozen pizzas is running a promotional special. Out of the next 100,000 boxes of pizza produced, randomly chosen boxes will be prize winners. There will be one grand prize winner who will receive $100,000. Five hundred first prize winners will get $1000, and 3,000 second prize winners will get a free pizza. Write the correct answer in fraction and percent form.

1. What is the probability that the box of pizza you just bought will be a grand prize winner?

____

2. What is the probability that the box of pizza you just bought will be a first prize winner?

____

3. What is the probability that the box of pizza you just bought will be a second prize winner?

____

4. What is the probability that you will win anything with the box of pizza you just bought?

____

Researchers at the National Institutes of Health are recommending that instead of screening all people for certain diseases, they can use a Punnett square to identify the people who are most likely to have the disease. By only screening these people, the cost of screening will be less. Fill in the Punnett square below and use them to choose the letter for the best answer.

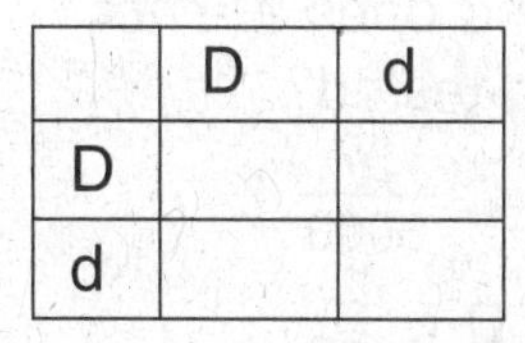

	D	d
D		
d		

5. What is the probability of DD?

A 0% **C** 50%

B 25% **D** 75%

6. What is the probability of Dd?

F 25% **H** 75%

G 50% **J** 100%

7. What is the probability of dd?

A 0% **C** 50%

B 25% **D** 75%

8. DD or Dd indicates that the patient will have the disease. What is the probability that the patient will have the disease?

F 25% **H** 75%

G 50% **J** 100%

Name ______________________ Date __________ Class __________

LESSON 10-5

Problem Solving

Independent and Dependent Events

Are the events independent or dependent? Write the correct answer.

1. Selecting a piece of fruit, then choosing a drink.

____

2. Buying a CD, then going to another store to buy a video tape if you have enough money left.

____

Dr. Fred Hoppe of McMaster University claims that the probability of winning a pick 6 number game where six numbers are drawn from the set 1 through 49 is about the same as getting 24 heads in a row when you flip a fair coin.

3. Find the probability of winning the pick 6 game and the probability of getting 24 heads in a row when you flip a fair coin.

____

____

4. Which is more likely: to win a pick 6 game or to get 24 heads in a row when you flip a fair coin?

____

____

In a shipment of 20 computers, 3 are defective. Choose the letter for the best answer.

5. Three computers are randomly selected and tested. What is the probability that all three are defective if the first and second ones are not replaced after being tested?

A $\frac{1}{760}$ C $\frac{27}{8000}$

B $\frac{1}{1140}$ D $\frac{3}{5000}$

6. Three computers are randomly selected and tested. What is the probability that all three are defective if the first and second ones are replaced after being tested?

F $\frac{1}{760}$ H $\frac{27}{8000}$

G $\frac{1}{1140}$ J $\frac{3}{5000}$

7. Three computers are randomly selected and tested. What is the probability that none are defective if the first and second ones are not replaced after being tested?

A $\frac{34}{57}$ C $\frac{4913}{6840}$

B $\frac{4913}{8000}$ D $\frac{1}{2000}$

8. Three computers are randomly selected and tested. What is the probability that none are defective if the first and second ones are replaced after being tested?

F $\frac{34}{57}$ H $\frac{4913}{6840}$

G $\frac{4913}{8000}$ J $\frac{1}{2000}$

Name ______________________ Date ___________ Class ___________

LESSON 10-6

Problem Solving

Making Decisions and Predictions

Write the correct answer.

1. A quality control inspector at a light bulb factory finds 2 defective bulbs in a batch of 1000 light bulbs. If the plant manufactures 75,000 light bulbs in one day, predict how many will be defective.

____

2. A game consists of rolling two fair number cubes labeled 1–6. Add both numbers. Player A wins if the sum is greater than 10. Player B wins if the sum is 7. Is the game fair or not? Explain.

____

3. A spinner has 5 equal sections numbered 1–5. Predict how many times Kevin will spin an even number in 40 spins.

____

4. In her last six 100-meter runs, Lee had the following times in seconds: 12:04, 13:11, 12:25, 11:58, 12:37, and 13:20. Based on these results, what is the best prediction of the number of times Lee will run faster than 13 seconds in her next 30 runs?

____

Use the table below that shows the number of colors of the last 200 T-shirts sold at a T-shirt shop. The manager of the store wants to order 1800 new T-shirts. Choose the letter of the best answer

T-Shirts Sold

Color	Number
Red	35
Blue	55
Green	15
Black	65
White	30

5. How many red T-shirts should the manager order?

   A 175 **C** 378

   B 315 **D** 630

6. How many blue T-shirts should the manager order?

   F 495 **H** 900

   G 665 **J** 990

7. How many more black T-shirts than white T-shirts should the manager order?

   A 855 **C** 315

   B 585 **D** 270

Name ______________________ Date ___________ Class ___________

LESSON **10-7**

Problem Solving
Odds

In the last 25 Summer Olympics since 1900, an American man has won the gold medal in the 400-meter dash 18 times. Write the correct answer.

1. Find the probability that an American man will win the gold medal in the 400-meter dash in the next Summer Olympics.

____

2. Find the probability that an American man will not win the gold medal in the 400-meter dash in the next Summer Olympics.

____

3. Find the odds that an American man will win the gold medal in the 400-meter dash in the next Summer Olympics.

____

4. Find the odds that an American man will not win the gold medal in the 400-meter dash in the next Summer Olympics.

____

Use the table below that shows the probability that a player will end up on a certain square after a single roll in a game of Monopoly.

Probability of Ending Up on a Monopoly Square

Square	Probability	Rank
In Jail	$\frac{39}{1000}$	1
Illinois Ave.	$\frac{32}{1000}$	2
Go	$\frac{31}{1000}$	3
Boardwalk	$\frac{26}{1000}$	18
Park Place	$\frac{22}{1000}$	33

5. What are the odds that you will end up in jail on your next roll in a game of Monopoly?

A 39:1000
B 39:961
C 1000:961
D 961:39

6. What are the odds that you will end up on Boardwalk on your next roll in a game of Monopoly?

A 13:500 **C** 13:487
B 500:13 **D** 487:13

7. What are the odds that you will not end up on Boardwalk on your next roll in a game of Monopoly?

F 487:500 **H** 13:487
G 500:487 **J** 487:13

8. What are the odds that you will end up on Go on your next roll in a game of Monopoly?

A 31:969 **C** 31:1000
B 969:31 **D** 1000:31

9. What are the odds that you will not end up on Park Place on your next roll in a game of Monopoly?

F 11:489 **H** 489:500
G 489:11 **J** 500:489

Name ______________________ Date __________ Class __________

LESSON 10-8

Problem Solving

Counting Principles

Write the correct answer.

1. The 5-digit zip code system for United States mail was implemented in 1963. How many different possibilities of zip codes are there with a 5-digit zip code where each digit can be 0 through 9?

____

2. In 1983, the ZIP +4 zip code system was introduced so mail could be more easily sorted by the 5-digit zip code plus an additional 4 digits. How many different possibilities of zip codes are there with the ZIP +4 system?

____

3. In Canada, each postal code has 6 symbols. The first, third and fifth symbols are letters of the alphabet and the second, fourth and sixth symbols are digits from 0 through 9. How many possible postal codes are there in Canada?

____

4. In the United Kingdom the postal code has 6 symbols. The first, second, fifth and sixth are letters of the alphabet and the third and fourth are digits from 0 through 9. How many possible postal codes are there in the United Kingdom?

____

Choose the letter for the best answer.

5. In Sharon Springs, Kansas, all of the phone numbers begin 852–4. The only differences in the phone numbers are the last 3 digits. How many possible phone numbers can be assigned using this system?

   A 729 **C** 6561
   B 1000 **D** 10,000

6. Many large cities have run out of phone numbers and so a new area code must be introduced. How many different phone numbers are there in a single area code if the first digit can't be zero?

   F 90,000 **H** 9,000,000
   G 4,782,969 **J** 10,000,000

7. How many different phone numbers are possible using a 3-digit area code and a 7-digit phone number if the first digit of the area code and phone number cannot be zero?

   A 3,486,784,401 **C** 9,500,000,000
   B 8,100,000,000 **D** 10,000,000,000

8. A shipping service offers to send packages by ground delivery using 2 different companies, by next day air using 3 different companies, and by 2-day air using 3 different companies. How many different shipping options does the service offer?

   F 3 **H** 10
   G 8 **J** 18

Name ____________________ Date __________ Class __________

LESSON 10-9

Problem Solving

Permutations and Combinations

Write the correct answer.

1. In a day camp, 6 children are picked to be team captains from the group of children numbered 1 through 49. How many possibilities are there for who could be the 6 captains?

____

2. If you had to match 6 players in the correct order for most popular outfielder from a pool of professional players numbered 1 through 49, how many possibilities are there?

____

Volleyball tournaments often use pool play to determine which teams will play in the semi-final and championship games. The teams are divided into different pools, and each team must play every other team in the pool. The teams with the best record in pool play advance to the final games.

3. If 12 teams are divided into 2 pools, how many games will be played in each pool?

____

4. If 12 teams are divided into 3 pools, how many pool play games will be played in each pool?

____

A word jumble game gives you a certain number of letters that you must make into a word. Choose the letter for the best answer.

5. How many possibilities are there for a jumble with 4 letters?

   A 4
   B 12
   C 24
   D 30

6. How many possibilities are there for a jumble with 5 letters?

   F 24
   G 75
   H 120
   J 150

7. How many possibilities are there for a jumble with 6 letters?

   A 120
   B 500
   C 720
   D 1000

8. On the Internet, a site offers a program that will un-jumble letters and give you all of the possible words that can be made with those letters. However, the program will not allow you to enter more than 7 letters due to the amount of time it would take to analyze. How many more possibilities are there with 8 letters than with 7?

   F 5040
   G 20,640
   H 35,280
   J 40,320

Name ______________________ Date ____________ Class ____________

LESSON **11-1**

Problem Solving

Simplifying Algebraic Expressions

Write the correct answer.

1. An item costs x dollars. The tax rate is 5% of the cost of the item, or $0.05x$. Write and simplify an expression to find the total cost of the item with tax.

   ____

2. A sweater costs d dollars at regular price. The sweater is reduced by 20%, or $0.2d$. Write and simplify an expression to find the cost of the sweater before tax.

   ____

3. Consecutive integers are integers that differ by one. You can represent consecutive integers as x, $x + 1$, $x + 2$ and so on. Write an equation and solve to find three consecutive integers whose sum is 33.

   ____

4. Consecutive even integers can be represented by x, $x + 2$, $x + 4$ and so on. Write an equation and solve to find three consecutive even integers whose sum is 54.

   ____

Choose the letter for the best answer.

5. In Super Bowl XXXV, the total number of points scored was 41. The winning team outscored the losing team by 27 points. What was the final score of the game?

   A 33 to 8
   B 34 to 7
   C 22 to 2
   D 18 to 6

6. A high school basketball court is 34 feet longer than it is wide. If the perimeter of the court is 268, what are the dimensions of the court?

   F 234 ft by 34 ft
   G 67 ft by 67 ft
   H 70 ft by 36 ft
   J 84 ft by 50 ft

7. Julia ordered 2 hamburgers and Steven ordered 3 hamburgers. If their total bill before tax was $7.50, how much did each hamburger cost?

   A $1.50
   B $1.25
   C $1.15
   D $1.02

8. On three tests, a student scored a total of 258 points. If the student improved his performance on each test by 5 points, what was the score on each test?

   F 81, 86, 91
   G 80, 85, 90
   H 75, 80, 85
   J 70, 75, 80

Name ______________________ Date __________ Class __________

LESSON 11-2

Problem Solving

Solving Multi-Step Equations

A taxi company charges $2.25 for the first mile and then $0.20 per mile for each mile after the first, or $F = \$2.25 + \$0.20(m - 1)$ where F is the fare and m is the number of miles.

1. If Juan's taxi fare was $6.05, how many miles did he travel in the taxi?

   ____

2. If Juan's taxi fare was $7.65, how many miles did he travel in the taxi?

   ____

A new car loses 20% of its original value when you buy it and then 8% of its original value per year, or $D = 0.8V - 0.08Vy$ where D is the value after y years with an original value V.

3. If a vehicle that was valued at $20,000 new is now worth $9,600, how old is the car?

   ____

4. A 6-year old vehicle is worth $12,000. What was the original value of the car?

   ____

The equation used to estimate typing speed is $S = \frac{1}{5}(w - 10e)$, where S is the accurate typing speed, w is the number of words typed in 5 minutes and e is the number of errors. Choose the letter of the best answer.

5. Jane can type 55 words per minute (wpm). In 5 minutes, she types 285 words. How many errors would you expect her to make?

   A 0 **C** 2
   B 1 **D** 5

6. If Alex types 300 words in 5 minutes with 5 errors, what is his typing speed?

   F 48 wpm **H** 59 wpm
   G 50 wpm **J** 60 wpm

7. Johanna receives a report that says her typing speed is 65 words per minute. She knows that she made 4 errors in the 5-minute test. How many words did she type in 5 minutes?

   A 285 **C** 365
   B 329 **D** 1825

8. Cecil can type 35 words per minute. In 5 minutes, she types 255 words. How many errors would you expect her to make?

   F 2 **H** 6
   G 4 **J** 8

Name ______________________ Date ____________ Class ____________

LESSON 11-3

Problem Solving

Solving Equations with Variables on Both Sides

The chart below describes three long-distance calling plans. Round to the nearest minute. Write the correct answer.

Long-Distance Plans

Plan	Monthly Access Fee	Charge per minute
A	$3.95	$0.08
B	$8.95	$0.06
C	$0	$0.10

1. For what number of minutes will plan A and plan B cost the same?

____

2. For what number of minutes per month will plan B and plan C cost the same?

____

3. For what number of minutes will plan A and plan C cost the same?

____

Choose the letter for the best answer.

4. Carpet Plus installs carpet for $100 plus $8 per square yard of carpet. Carpet World charges $75 for installation and $10 per square yard of carpet. Find the number of square yards of carpet for which the cost including carpet and installation is the same.

   A 1.4 yd^2 **C** 12.5 yd^2
   B 9.7 yd^2 **D** 87.5 yd^2

5. One shuttle service charges $10 for pickup and $0.10 per mile. The other shuttle service has no pickup fee but charges $0.35 per mile. Find the number of miles for which the cost of the shuttle services is the same.

   F 2.5 miles
   G 22 miles
   H 40 miles
   J 48 miles

6. Joshua can purchase tile at one store for $0.99 per tile, but he will have to rent a tile saw for $25. At another store he can buy tile for $1.50 per tile and borrow a tile saw for free. Find the number of tiles for which the cost is the same. Round to the nearest tile.

   A 10 tiles **C** 25 tiles
   B 13 tiles **D** 49 tiles

7. One plumber charges a fee of $75 per service call plus $15 per hour. Another plumber has no flat fee, but charges $25 per hour. Find the number of hours for which the cost of the two plumbers is the same.

   F 2.1 hours **H** 7.5 hours
   G 7 hours **J** 7.8 hours

Name ______________________ Date __________ Class __________

LESSON 11-4

Problem Solving

Solving Inequalities by Multiplying or Dividing

Write the correct answer

1. A bottle contains at least 4 times as much juice as a glass contains. The bottle contains 32 fluid ounces. Write an inequality that shows this relationship.

____

2. Solve the inequality in Exercise 1. What is the greatest amount the glass could contain?

____

3. In the triple jump, Katrina jumped less than one-third the distance that Paula jumped. Katrina jumped 5 ft 6 in. Write an inequality that shows this relationship.

____

4. Solve the inequality in Exercise 3. How far could Paula could have jumped?

____

____

Choose the letter for the best answer.

5. Melinda earned at least 3 times as much money this month as last month. She earned $567 this month. Which inequality shows this relationship?

A $567 < x$

B $567 < 3x$

C $567 > 3x$

D $567 \geq 3x$

6. The shallow end of a pool is less than one-quarter as deep as the deep end. The shallow end is 3 feet deep. Which inequality shows this relationship?

F $4 > 3x$

G $4x < 3$

H $\frac{x}{4} > 3$

J $\frac{x}{4} < 3$

7. Arthur worked in the garden more than half as long as his brother. Arthur worked 6 hours in the garden. How long did his brother work in the garden?

A less than 3 hours

B 3 hours

C less than 12 hours

D more than 12 hours

8. The distance from Bill's house to the library is no more than 5 times the distance from his house to the park. If Bill's house is 10 miles from the library, what is the greatest distance his house could be from the park?

F 2 miles

G more than 2 miles

H 20 miles

J less than 20 miles

Name ______________________ Date __________ Class __________

LESSON 11-5

Problem Solving

Solving Two-Step Inequalities

A school club is selling printed T-shirts to raise $650 for a trip. The table shows the profit they will make on each shirt after they pay the cost of production.

Shirt	Profit
50/50	$5.50
100% cotton	$7.82

1. Suppose the club already has $150, at least how many 50/50 shirts must they sell to make enough money for the trip?

____

2. Suppose the club already has $100, but it plans to spend $50 on advertising. At least how many 100% cotton shirts must they sell to make enough money for the trip?

____

3. Suppose the club sold thirty 50/50 shirts on the first day of sales. At least how many more 50/50 shirts must they sell to make enough money for the trip?

____

For Exercises 4–5, use this equation to estimate typing speed, $S = \frac{w}{5} - 2e$, where S is the accurate typing speed, w is the number of words typed in 5 minutes, and e is the number of errors. Choose the letter for the best answer.

4. One of the qualifications for a job is a typing speed of at least 65 words per minute. If Jordan knows that she will be able to type 350 words in five minutes, what is the maximum number of errors she can make?

A 0 **C** 3

B 2 **D** 4

5. Tanner usually makes 3 errors every 5 minutes when he is typing. If his goal is an accurate typing speed of at least 55 words per minute, how many words does he have to be able to type in 5 minutes?

F 61 words **H** 305 words

G 300 words **J** 325 words

6. A taxi charges $2.05 per ride and $0.20 for each mile, which can be written as $F = \$2.05 + \$0.20m$. How many miles can you travel in the cab and have the fare be less than $10?

A 15 **C** 39

B 25 **D** 43

7. Celia's long distance company charges $5.95 per month plus $0.06 per minute. If Celia has budgeted $30 for long distance, what is the maximum number of minutes she can call long distance per month?

F 375 minutes **H** 405 minutes

G 400 minutes **J** 420 minutes

Name ______________________________ Date ____________ Class ____________

LESSON 11-6

Problem Solving

Systems of Equations

After college, Julia is offered two different jobs. The table summarizes the pay offered with each job. Write the correct answer.

Job	Yearly Salary	Yearly Increase
A	$20,000	$2500
B	$25,000	$2000

1. Write an equation that shows the pay y of Job A after x years.

____

2. Write an equation that shows the pay y of Job B after x years.

____

3. Is (8, 35,000) a solution to the system of equations in Exercises 1 and 2?

____

4. Solve the system of equations in Exercises 1 and 2.

____

5. If Julia plans to stay at this job only a few years and pay is the only consideration, which job should she choose?

____

A travel agency is offering two Orlando trip plans that include hotel accommodations and pairs of tickets to theme parks. Use the table below. Choose the letter for the best answer.

Trip	Number of nights	Pairs of theme park tickets	Cost
A	3	2	$415
B	5	4	$725

6. Find an equation about trip A where x represents the hotel cost per night and y represents the cost per pair of theme park tickets.

A $5x + 2y = 415$ **C** $8x + 6y = 415$

B $2x + 3y = 415$ **D** $3x + 2y = 415$

7. Find an equation about trip B where x represents the hotel cost per night and y represents the cost per pair of theme park tickets.

F $5x + 4y = 725$

G $4x + 5y = 725$

H $8x + 6y = 725$

J $3x + 4y = 725$

8. Solve the system of equations to find the nightly hotel cost and the cost for each pair of theme park tickets.

A ($50, $105)

B ($125 $20)

C ($105, $50)

D ($115, $35)

Name ______________________ Date ____________ Class ____________

LESSON 12-1

Problem Solving

Graphing Linear Equations

Write the correct answer.

1. The distance in feet traveled by a falling object is found by the formula $d = 16t^2$ where d is the distance in feet and t is the time in seconds. Graph the equation. Is the equation linear?

____

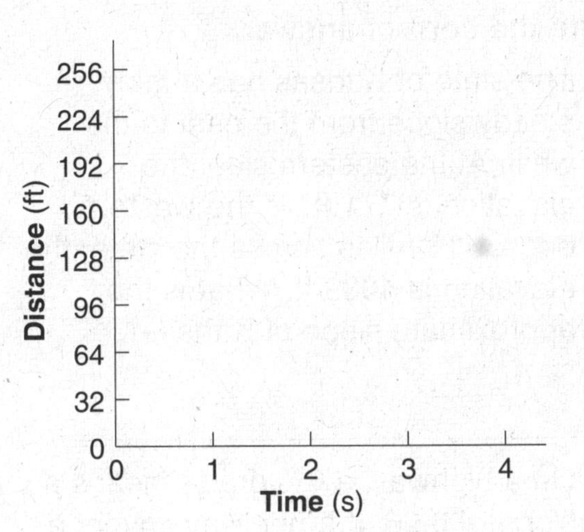

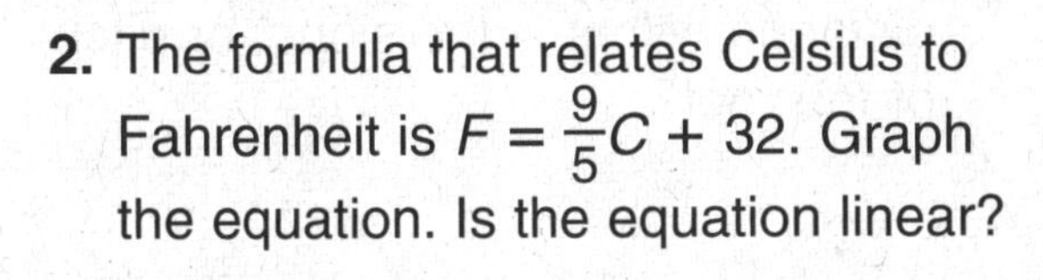

2. The formula that relates Celsius to Fahrenheit is $F = \frac{9}{5}C + 32$. Graph the equation. Is the equation linear?

____

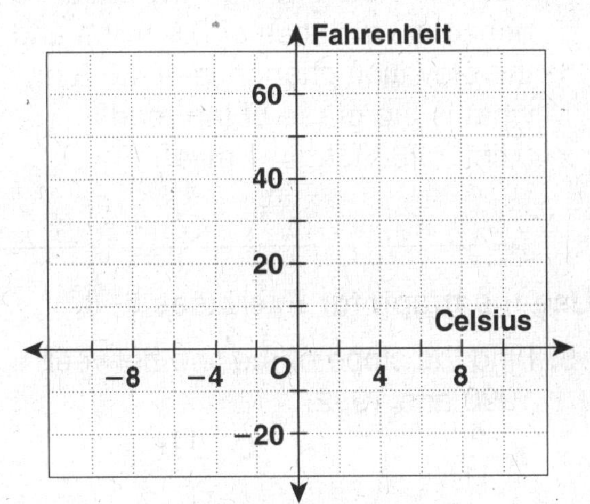

Wind chill is the temperature that the air feels like with the effect of the wind. The graph below shows the wind chill equation for a wind speed of 25 mph. For Exercises 3–6, refer to the graph.

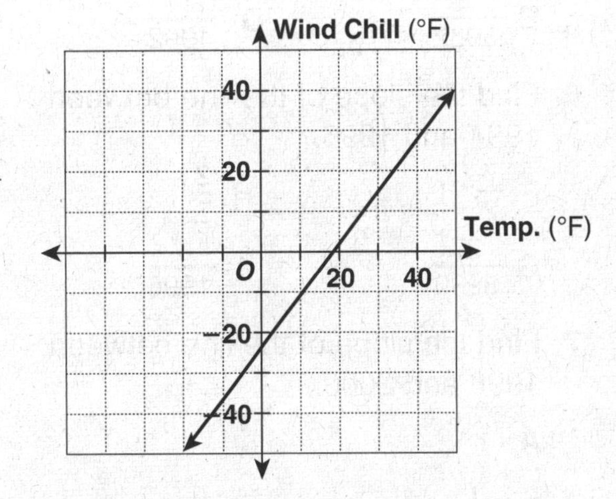

3. If the temperature is 40° with a 25 mph wind, what is the wind chill?

   A 6° **C** 29°

   B 20° **D** 40°

4. If the temperature is 20° with a 25 mph wind, what is the wind chill?

   F 3° **H** 13°

   G 10° **J** 20°

5. If the temperature is 0° with a 25 mph wind, what is the wind chill?

   A −30° **C** −15°

   B −24° **D** 0°

6. If the wind chill is 10° and there is a 25 mph wind, what is the actual temperature?

   F −11° **H** 15°

   G 0° **J** 25°

Name ______________________ Date __________ Class __________

LESSON **12-2**

Problem Solving

Slope of a Line

Write the correct answer.

1. The state of Kansas has a fairly steady slope from the east to the west. At the eastern side, the elevation is 771 ft. At the western edge, 413 miles across the state, the elevation is 4039 ft. What is the approximate slope of Kansas?

____

2. The Feathered Serpent Pyramid in Teotihuacan, Mexico, has a square base. From the center of the base to the center of an edge of the pyramid is 32.5 m. The pyramid is 19.4 m high. What is the slope of each face of the pyramid?

____

3. On a highway, a 6% grade means a slope of 0.06. If a highway covers a horizontal distance of 0.5 miles and the elevation change is 184.8 feet, what is the grade of the road? (Hint: 5280 feet = 1 mile.)

____

4. The roof of a house rises vertically 3 feet for every 12 feet of horizontal distance. What is the slope, or pitch of the roof?

____

Use the graph for Exercises 5–8.

Number of Earthquakes Worldwide with a Magnitude of 7.0 or Greater

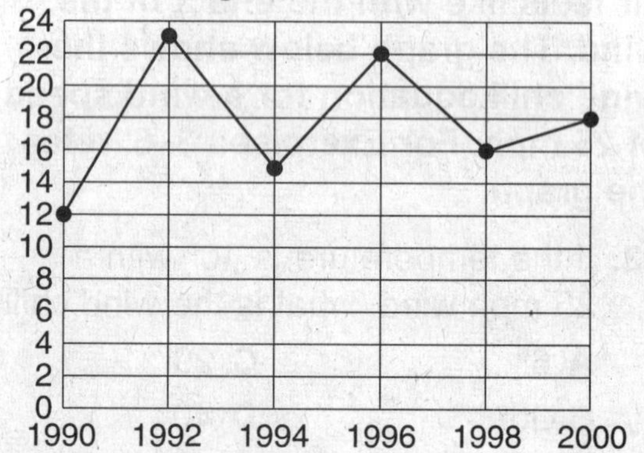

5. Find the slope of the line between 1990 and 1992.

   A $\frac{2}{11}$ **C** $\frac{11}{2}$

   B $\frac{35}{3982}$ **D** $\frac{11}{1992}$

6. Find the slope of the line between 1994 and 1996.

   F $\frac{7}{2}$ **H** $\frac{2}{7}$

   G $\frac{37}{3990}$ **J** $\frac{7}{1996}$

7. Find the slope of the line between 1998 and 2000.

   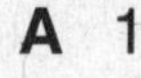

   A 1

   B $\frac{1}{999}$

   C $\frac{1}{1000}$

   D 2

8. What does it mean when the slope is negative?

   F The number of earthquakes stayed the same.

   G The number of earthquakes increased.

   H The number of earthquakes decreased.

   J It means nothing.

Name ______________________ Date __________ Class __________

LESSON 12-3

Problem Solving

Using Slopes and Intercepts

Write the correct answer.

1. Jaime purchased a $20 bus pass. Each time she rides the bus, $1.25 is deducted from the pass. The linear equation $y = -1.25x + 20$ represents the amount of money on the bus pass after x rides. Identify the slope and the x- and y-intercepts. Graph the equation at the right.

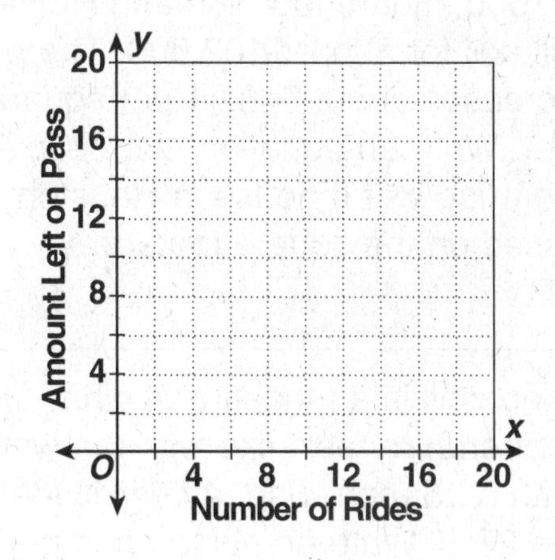

____

____

2. The rent charged for space in an office building is related to the size of the space rented. The rent for 600 square feet of floor space is $750, while the rent for 900 square feet is $1150. Write an equation for the rent y based on the square footage of the floor space x.

____

Choose the letter of the correct answer.

3. A limousine charges $35 plus $2 per mile. Which equation shows the total cost of a ride in the limousine?

   A $y = 35x + 2$ **C** $y = 2x - 35$
   B $y = 2x + 35$ **D** $2x + 35y = 2$

4. A newspaper pays its delivery people $75 each day plus $0.10 per paper delivered. Which equation shows the daily earnings of a delivery person?

   F $y = 0.1x + 75$ **H** $x + 0.1y = 75$
   G $y = 75x + 0.1$ **J** $0.1x + y = 75$

5. A friend gave Ms. Morris a $50 gift card for a local car wash. If each car wash costs $6, which equation shows the number of dollars left on the card?

   A $50x + 6y = 1$ **C** $y = -6x + 50$
   B $y = 6x + 50$ **D** $y = 6x - 50$

6. Antonio's weekly allowance is given by the equation $A = 0.5c + 10$, where c is the number of chores he does. If he received $16 in allowance one week, how many chores did he do?

   F 10 **H** 14
   G 12 **J** 15

Name ______________________ Date __________ Class __________

LESSON **12-4**

Problem Solving

Point-Slope Form

Write the correct answer.

1. A 1600 square foot home in Houston will sell for about $102,000. The price increases about $43.41 per square foot. Write an equation that describes the price y of a house in Houston, based on the square footage x.

____

2. Write the equation in Exercise 1 in slope-intercept form.

____

3. Wind chill is a measure of what temperature feels like with the wind. With a 25 mph wind, 40°F will feel like 29°F. Write an equation in point-slope form that describes the wind chill y based on the temperature x, if the slope of the line is 1.337.

____

4. With a 25 mph wind, what does a temperature of 0°F feel like?

____

From 2 to 13 years, the growth rate for children is generally linear. Choose the letter of the correct answer.

5. The average height of a 2-year old boy is 36 inches, and the average growth rate per year is 2.2 inches. Write an equation in point-slope form that describes the height of a boy y based on his age x.

A $y - 36 = 2(x - 2.2)$

B $y - 2 = 2.2(x - 36)$

C $y - 36 = 2.2(x - 2)$

D $y - 2.2 = 2(x - 36)$

6. The average height of a 5-year old girl is 44 inches, and the average growth rate per year is 2.4 inches. Write an equation in point-slope form that describes the height of a girl y based on her age x.

F $y - 2.4 = 44(x - 5)$

G $y - 44 = 2.4(x - 5)$

H $y - 44 = 5(x - 2.4)$

J $y - 5 = 2.4(x - 44)$

7. Write the equation from Exercise 6 in slope-intercept form.

A $y = 2.4x - 100.6$

B $y = 44x - 217.6$

C $y = 5x + 32$

D $y = 2.4x + 32$

8. Use the equation in Exercise 6 to find the average height of a 13-year old girl.

F 56.3 in.

G 63.2 in.

H 69.4 in.

J 97 in.

Name ______________________ Date __________ Class __________

LESSON 12-5

Problem Solving

Direct Variation

Determine whether the data sets show direct variation. If so, find the equation of direct variation.

1. The table shows the distance in feet traveled by a falling object in certain times.

Time (s)	0	0.5	1	1.5	2	2.5	3
Distance (ft)	0	4	16	36	64	100	144

____

____

2. The R-value of insulation gives the material's resistance to heat flow. The table shows the R-value for different thicknesses of fiberglass insulation.

Thickness (in)	1	2	3	4	5	6
R-value	3.14	6.28	9.42	12.56	15.7	18.84

____

____

3. The table shows the lifting power of hot air.

Hot Air (ft^3)	50	100	500	1000	2000	3000
Lift (lb)	1	2	10	20	40	60

____

____

4. The table shows the relationship between degrees Celsius and degrees Fahrenheit.

° **Celsius**	−10	−5	0	5	10	20	30
° **Fahrenheit**	14	23	32	41	50	68	86

____

____

The relationship between your weight on Earth and your weight on other planets is direct variation. The table below shows how much a person who weights 100 lb on Earth would weigh on the moon and different planets.

Solar System Objects	**Weight** (lb)
Moon	16.6
Jupiter	236.4
Pluto	6.7

5. Find the equation of direct variation for the weight on earth e and on the moon m.

A $m = 0.166e$

B $m = 16.6e$

C $m = 6.02e$

D $m = 1660e$

6. How much would a 150 lb person weigh on Jupiter?

F 63.5 lb

G 286.4 lb

H 354.6 lb

J 483.7 lb

7. How much would a 150 lb person weigh on Pluto?

A 5.8 lb

B 10.05 lb

C 12.3 lb

D 2238.8 lb

Name ________________________________ Date ____________ Class ____________

LESSON **12-6**

Problem Solving

Graphing Inequalities in Two Variables

The senior class is raising money by selling popcorn and soft drinks. They make $0.25 profit on each soft drink sold, and $0.50 on each bag of popcorn. Their goal is to make at least $500.

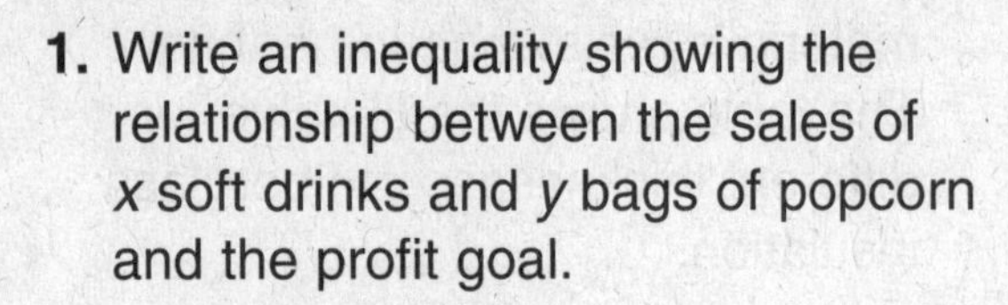

1. Write an inequality showing the relationship between the sales of x soft drinks and y bags of popcorn and the profit goal.

____

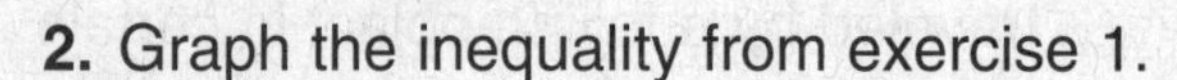

2. Graph the inequality from exercise 1.

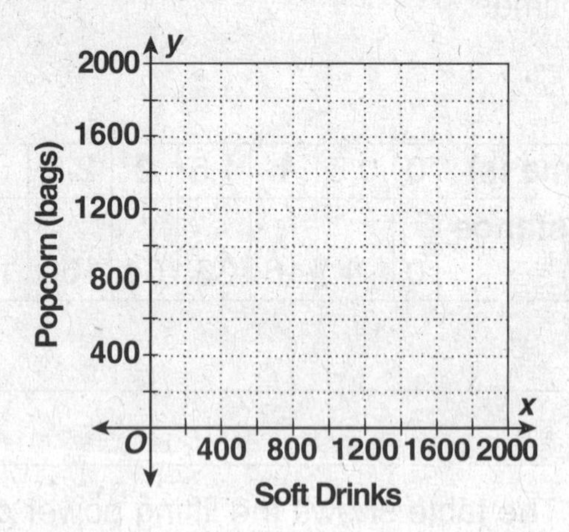

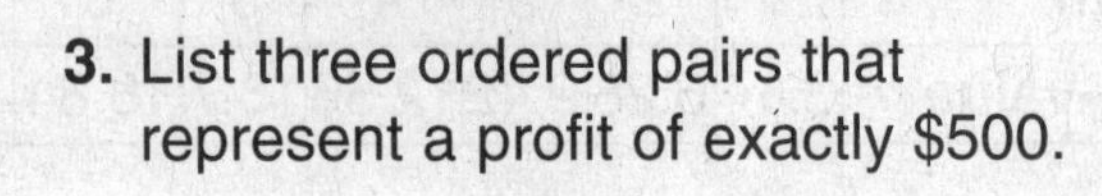

3. List three ordered pairs that represent a profit of exactly $500.

____

____

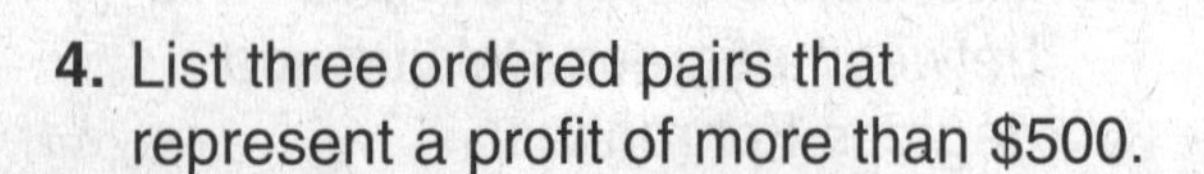

4. List three ordered pairs that represent a profit of more than $500.

____

____

5. List three ordered pairs that represent a profit of less than $500.

____

____

A vehicle is rated to get 19 mpg in the city and 25 mpg on the highway. The vehicle has a 15-gallon gas tank. The graph below shows the number of miles you can drive using no more than 15 gallons.

6. Write the inequality represented by the graph.

A $\frac{x}{19} + \frac{y}{25} < 15$

B $\frac{x}{19} + \frac{y}{25} \leq 15$

C $\frac{x}{19} + \frac{y}{25} \geq 15$

D $\frac{x}{19} + \frac{y}{25} > 15$

7. Which ordered pair represents city and highway miles that you can drive on one tank of gas?

F (200, 150) **H** (250, 75)

G (50, 350) **J** (100, 175)

8. Which ordered pair represents city and highway miles that you cannot drive on one tank of gas?

A (100, 200) **C** (50, 275)

B (150, 200) **D** (250, 25)

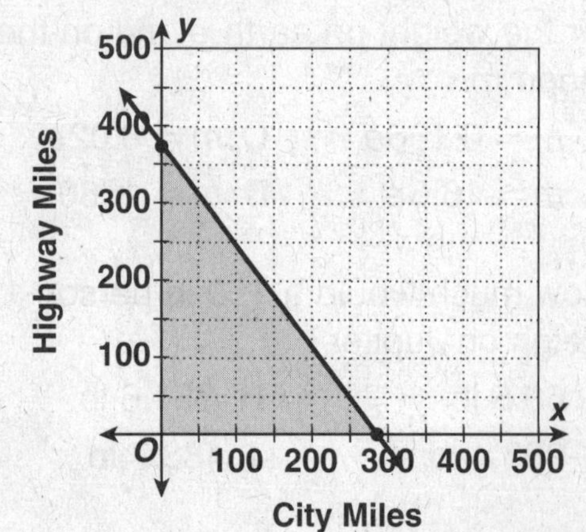

Name ______________________ Date __________ Class __________

LESSON 12-7

Problem Solving

Lines of Best Fit

Write the correct answer. Round to the nearest hundredth.

1. The table shows in what year different average speed barriers were broken at the Indianapolis 500. If x is the year, with $x = 0$ representing 1900, and y is the average speed, find the mean of the x- and y-coordinates.

Barrier (mi/h)	Year	Average Speed (mi/h)
80	1914	82.5
100	1925	101.1
120	1949	121.3
140	1962	140.9
160	1972	163.0
180	1990	186.0

____

2. Graph the data from exercise 1 and find the equation of the line of best fit.

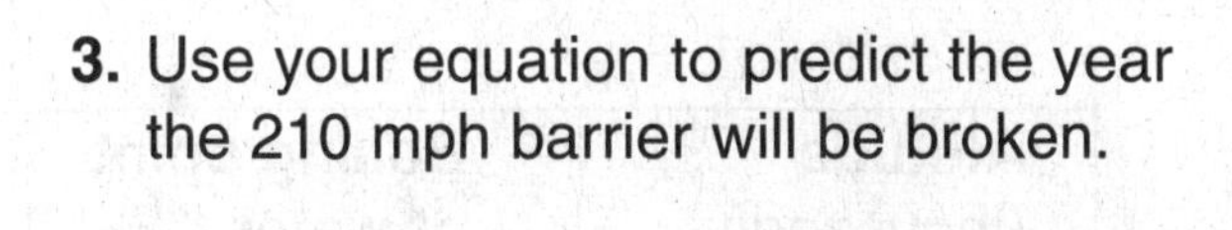

____

____

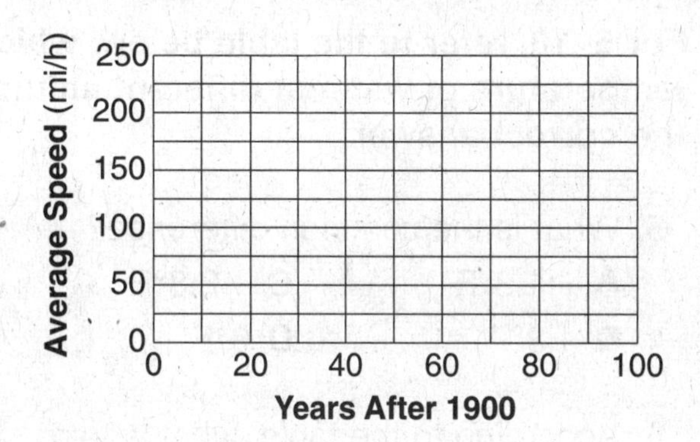

3. Use your equation to predict the year the 210 mph barrier will be broken.

____

The percent of the U.S. population who smokes can be represented by the line of best fit with the equation $y = -0.57x + 44.51$ where x is the year, $x = 0$ represents 1960, and y is the percent of the population who smokes. Circle the letter of the correct answer.

4. Which term describes the percent of the population that smokes?

   A Increasing **C** No change
   B Decreasing **D** Cannot tell

5. Use the equation to predict the percent of smokers in 2005.

   F 13.16% **H** 24.56%
   G 18.86% **J** 41.66%

6. Use the equation to predict when the percent of smokers will be less than 15%.

   A 1996 **C** 2012
   B 2010 **D** 2023

7. Use the equation to predict the percent of smokers in 2010.

   F 10.31% **H** 16.01%
   G 11.01% **J** 12.71%

Name ______________________ Date __________ Class __________

LESSON 13-1

Problem Solving

Terms of Arithmetic Sequences

A section of seats in an auditorium has 18 seats in the first row. Each row has two more seats than the previous row. There are 25 rows in the section. Write the correct answer.

1. List the number of seats in the second, third and fourth rows of the section.

   ____

2. How many seats are in the 10th row?

   ____

3. How many seats are in the 15th row?

   ____

4. In which row are there 32 seats?

   ____

For 5–10, refer to the table below, which shows the boiling temperature of water at different altitudes. Choose the letter of the correct answer.

Altitude (thousands of feet)	**Boiling point of water** (°F)
1	210.2
2	208.4
3	206.6
4	204.8
5	203

5. What is the common difference?

   A −1.8°F **C** −2.8°F
   B 1.8°F **D** 6°F

6. According to the table, what would be the boiling point of water at an altitude of 10,000 feet?

   F 192.2°F **H** 226.4°F
   G 194°F **J** 228.2°F

7. According to the table, what would be the boiling point of water at an altitude of 15,000 feet?

   A 181.4°F **C** 185°F
   B 183.2°F **D** 235.4°F

8. Estimate the boiling point of water in Jacksonville, Florida, which has an elevation of 0 feet.

   F 0°F **H** 212°F
   G 208.4°F **J** 213.8°F

9. The highest point in the United States is Mt. McKinley, Alaska, with an elevation of 20,320 feet. Estimate the boiling point of water at the top of Mt. McKinley.

   A 172.4°F **C** 244.4°F
   B 176°F **D** 246.2°F

10. At which elevation will the boiling point of water be less than 150°F?

    F 28,000 ft **H** 32,000 ft
    G 30,000 ft **J** 35,000 ft

Name ______________________ Date __________ Class __________

LESSON 13-2

Problem Solving

Terms of Geometric Sequences

For Exercises 1–2, determine if the sequence could be geometric. If so, find the common ratio. Write the correct answer.

1. A computer that was worth $1000 when purchased was worth $800 after six months, $640 after a year, $512 after 18 months, and $409.60 after two years.

____

2. A student works for a starting wage of $6.00 per hour. She is told that she can expect a $0.25 raise every six months.

____

3. A piece of paper that is 0.01 inches thick is folded in half repeatedly. If the paper were folded 6 times, how thick would the result be?

____

4. A vacuum pump removes one-half of the air in a container with each stroke. How much of the original air is left in the container after 8 strokes?

____

For exercises 5–8, assume that the cost of a college education increases an average of 5% per year. Choose the letter of the correct answer.

5. If the in-state tuition at the University of Florida is $2256 per year, what will the tuition be in 10 years?

   A $3174.24
   B $3333.14
   C $3499.80
   D $3674.79

6. If it costs $3046 per year for tuition for a Virginia resident at the University of Virginia now, how much will tuition be in 8 years?

   F $4183.26
   G $4286.03
   H $4500.33
   J $4725.35

7. If it costs $25,839 per year in tuition to attend Northwestern University now, how much will tuition be in 5 years?

   A $31,407.47
   B $32,977.84
   C $37,965.97
   D $42,483.72

8. If you start attending Northwestern University in 5 years and attend for 4 years, how much will you spend in total for tuition?

   F $142,138.61
   G $135,370.12
   H $131,911.36
   J $169,934.88

Name ________________________________ Date ____________ Class ____________

LESSON **13-3**

Problem Solving

Other Sequences

A toy rocket is launched and the height of the rocket during its first four seconds is recorded. Write the correct answer.

Time (sec)	**Height** (ft)
0	0
1	176
2	320
3	432
4	512
5	
6	
7	

1. Find the first differences for the rocket's heights.

____

2. Find the second differences.

____

3. Use the first and second differences to predict the height of the rocket at 5, 6, and 7 seconds.

4. What is the maximum height of the rocket?

____

5. When will the rocket hit the ground?

____

For exercises 6–9, refer to the table below, which shows the number of diagonals for different polygons. Choose the letter for the correct answer.

Polygon	**Sides**	**Diagonals**
Triangle	3	0
Quadrilateral	4	2
Pentagon	5	5
Hexagon	6	9
Heptagon	7	14

6. What are the first differences for the diagonals?

A 1, 1, 1, 1
B 3, 2, 0, 3, 7
C 2, 3, 4, 5
D 2, 7, 14, 23

7. What are the second differences?

F 1, 1, 1
G 1, 2, 3, 4
H 5, 7, 8
J 7, 9, 11, 13

8. How many diagonals does a nonagon (9 sides) have?

A 21
B 24
C 27
D 32

9. Which rule will give the number of diagonals d for s sides?

F $d = \frac{s(s+1)}{2}$

G $d = (s-3)(s-2) - 1$

H $d = \frac{s(s-3)}{2}$

J $d = (s-3)(s-2)$

Name ______________________ Date __________ Class __________

LESSON 13-4

Problem Solving

Linear Functions

Write the correct answer.

1. The greatest amount of snow that has ever fallen in a 24-hour period in North America was on April 14–15, 1921 in Silver Lake, Colorado. In 24 hours, 76 inches of snow fell, at an average rate of 3.2 inches per hour. Find a rule for the linear function that describes the amount of snow after x hours at the average rate.

____

2. At the average rate of snowfall from Exercise 1, how much snow had fallen in 15 hours?

____

3. The altitude of clouds in feet can be found by multiplying the difference between the temperature and the dew point by 228. If the temperature is 75°, find a rule for the linear function that describes the height of the clouds with dew point x.

____

4. If the temperature is 75° and the dew point is 40°, what is the height of the clouds?

____

For exercises 5–7, refer to the table below, which shows the relationship between the number of times a cricket chirps in a minute and temperature.

Cricket Chirps/min	Temperature (°F)
80	60
100	65
120	70
140	75

5. Find a rule for the linear function that describes the temperature based on x, the number of cricket chirps in a minute based on temperature.

   A $f(x) = x + 5$

   B $f(x) = \frac{x}{4} + 40$

   C $f(x) = x - 20$

   D $f(x) = \frac{x}{2} + 20$

6. What is the temperature if a cricket chirps 150 times in a minute?

   F 77.5°F

   G 95°F

   H 130°F

   J 155°F

7. If the temperature is 85°F, how many times will a cricket chirp in a minute?

   A 61

   B 105

   C 180

   D 200

Name ______________________ Date __________ Class __________

LESSON 13-5

Problem Solving

Exponential Functions

From 1950 to 2000, the world's population grew exponentially. The function that models the growth is $f(x) = 1.056 \cdot 1.018^x$ where x is the year ($x = 50$ represents 1950) and $f(x)$ is the population in billions. Round each number to the nearest hundredth.

1. Estimate the world's population in 1950.

____

2. Estimate the world's population in 2005.

____

3. Predict the world's population in 2025.

____

4. Predict the world's population in 2050.

____

Insulin is used to treat people with diabetes. The table below shows the percent of an insulin dose left in the body at different times after injection.

Time elapsed (min)	**Percent remaining**
0	100
48	50
96	25
144	12.5

5. Which ordered pair does not represent a half-life of insulin?

A (24, 70.71) **C** (48, 50)

B (50, 50) **D** (72, 35.35)

6. Write an exponential function that describes the percent of insulin in the body after x half-lives.

F $f(x) = 100\left(\frac{1}{2}\right)^x$ **H** $f(x) = 2(100)^x$

G $f(x) = 10\left(\frac{1}{2}\right)^x$ **J** $f(x) = 48\left(\frac{1}{2}\right)^x$

7. What percent of insulin would be left in the body after 6 hours?

A 0.25% **C** 0.55%

B 0.39% **D** 1.56%

8. What percent of insulin would be left in the body after 9 hours?

F 0.04% **H** 0.17%

G 0.12% **J** 0.26%

9. A new form of insulin that is being developed has a half-life of 9 hours. Write an exponential function that describes the percent of insulin in the body after x half-lives.

A $f(x) = 100\left(\frac{1}{2}\right)^x$ **C** $f(x) = 2(100)^x$

B $f(x) = 9\left(\frac{1}{2}\right)^x$ **D** $f(x) = 100(9)^x$

10. What percent of the new form of insulin would be left in the body after 9 hours?

F 12.5% **H** 50%

G 25% **J** 75%

Name ______________________ Date __________ Class __________

LESSON 13-6

Problem Solving

Quadratic Functions

To find the time it takes an object to fall, you can use the equation $h = -16t^2 - vt + s$ where h is the height in feet, t is the time in seconds, v is the initial velocity, and s is the starting height in feet. Write the correct answer.

1. If a construction worker drops a tool from 240 feet above the ground, how many feet above the ground will it be in 2 seconds? Hint: $v = 0$, $s = 240$.

   ____

2. How long will it take the tool in Exercise 1 to hit the ground? Round to the nearest hundredth.

   ____

3. The Gateway Arch in St. Louis, Missouri is the tallest manmade memorial. The arch rises to a height of 630 feet. If you throw a rock down from the top of the arch with a velocity of 20 ft/s, how many feet above the ground will the rock be in 2 seconds?

   ____

4. Will the rock in exercise 3 hit the ground within 6 seconds of throwing it?

   ____

The average monthly rainfall for Seattle, Washington can be approximated by the equation $f(x) = 0.147x^2 - 1.890x + 7.139$ where x is the month (January: $x = 1$, February, $x = 2$, etc.) and $f(x)$ is the monthly rainfall in inches. Choose the letter for the best answer.

5. What is the average monthly rainfall in Seattle for the month of January?
   - **A** 3.7 in
   - **B** 5.4 in
   - **C** 7.6 in
   - **D** 9.2 in

6. What is the average monthly rainfall in Seattle for the month of April?
   - **F** 0.2 in
   - **G** 1.4 in
   - **H** 1.9 in
   - **J** 2.8 in

7. What is the average monthly rainfall in Seattle for the month of August?
   - **A** 1.1 in
   - **B** 1.4 in
   - **C** 5.6 in
   - **D** 6.8 in

8. In what month does it rain the least in Seattle, Washington?
   - **F** May
   - **G** June
   - **H** July
   - **J** August

Name ______________________ Date ____________ Class ____________

LESSON 13-7

Problem Solving

Inverse Variation

For a given focal length of a camera, the f-stop varies inversely with the diameter of the lens. The table below gives the f-stop and diameter data for a focal length of 400 mm. Round to the nearest hundredth.

f-stop	diameter (mm)
1	400
2	200
4	100
8	50
16	25
32	12.5

1. Use the table to write an inverse variation function.

   ____

2. What is the diameter of a lens with an f-stop of 1.4?

   ____

3. What is the diameter of a lens with an f-stop of 11?

   ____

4. What is the diameter of a lens with an f-stop of 22?

   ____

The inverse square law of radiation says that the intensity of illumination varies inversely with the square of the distance to the light source.

5. Using the inverse square law of radiation, if you halve the distance between yourself and a fire, by how much will you increase the heat you feel?

   A 2

   B 4

   C 8

   D 16

6. Using the inverse square law of radiation, if you double the distance between a radio and the transmitter, how will it affect the signal intensity?

   F $\frac{1}{4}$ as strong

   G $\frac{1}{2}$ as strong

   H twice as strong

   J 4 times stronger

7. Using the inverse square law of radiation, if you increase the distance between yourself and a light by 4 times, how will it affect the light's intensity?

   A $\frac{1}{16}$ as strong

   B $\frac{1}{4}$ as strong

   C $\frac{1}{2}$ as strong

   D twice as strong

8. Using the inverse square law of radiation if you move 3 times closer to a fire, how much more intense will the fire feel?

   F $\frac{1}{3}$ as strong

   G 3 times stronger

   H 9 times stronger

   J 27 times stronger

Name ______________________ Date __________ Class __________

LESSON 14-1

Problem Solving

Polynomials

The table below shows expressions used to calculate the surface area and volume of various solid figures where *s* is side length, *l* is length, *w* is width, *h* is height, and *r* is radius.

Solid Figure Polynomials

Solid Figure	Surface Area	Volume
Cube	$6s^2$	s^3
Rectangular Prism	$2lw + 2lh + 2wh$	lwh
Right Cone	$\pi rl + \pi r^2$	$\pi r^2 h$
Sphere	$4\pi r^2$	$\frac{4}{3}\pi r^3$

1. List the expressions that are trinomials.

____

2. What is the degree of the expression for the surface area of a sphere?

____

3. A cube has side length of 5 inches. What is its surface area?

____

4. If you know the radius and height of a cone, you can use the expression $(r^2 + h^2)^{0.5}$ to find its slant height. Is this expression a polynomial? Why or why not?

____

____

5. If a sphere has a radius of 4 feet, what is its surface area and volume? Use $\frac{22}{7}$ for pi.

____

Circle the letter of the correct answer.

6. Which statement is true of all the polynomials in the volume column of the table?

   A They are trinomials

   B They are binomials.

   C They are monomials.

   D None of them are polynomials.

7. The height, in feet, of a baseball thrown straight up into the air from 6 feet above the ground at 100 feet per second after *t* seconds is given by the polynomial $-16t^2 + 100t + 6$. What is the height of the baseball 4 seconds after it was thrown?

   F 150 feet

   G 278 feet

   H 342 feet

   J 662 feet

Name ____________________ Date __________ Class __________

LESSON 14-2

Problem Solving

Simplifying Polynomials

Write the correct answer.

1. The area of a trapezoid can be found using the expression $\frac{h}{2}(b_1 + b_2)$ where h is height, b is the length of $base_1$, and b_2 is the length of $base_2$. Use the Distributive Property to write an equivalent expression.

____

2. The sum of the measures of the interior angles of a polygon with n sides is $180(n - 2)$ degrees. Use the Distributive Property to write an equivalent expression, and use the expression to find the sum of the measures of the interior angles of an octagon.

____

3. The volume of a box of height h is $2h^4 + h^3 + h^2 + h^2 + h$ cubic inches. Simplify the polynomial and then find the volume if the height of the box is 3 inches.

____

____

4. The height, in feet, of a rocket launched upward from the ground with an initial velocity of 64 feet per second after t seconds is given by $16(4t - t^2)$. Write an equivalent expression for the rocket's height after t seconds. What is the height of the rocket after 4 seconds?

____

____

Circle the letter of the correct answer.

5. The surface area of a square pyramid with base b and slant height l is given by the expression $b(b + 2l)$. What is the surface area of a square pyramid with base 3 inches and slant height 5 inches?

A 13 square inches

B 19 square inches

C 39 square inches

D 55 square inches

6. The volume of a box with a width of $3x$, a height of $4x - 2$, and a length of $3x + 5$ can be found using the expression $3x(12x^2 + 14x - 10)$. Which is this expression, simplified by using the Distributive Property?

F $36x^2 + 42x - 30$

G $15x^3 + 17x^2 - 7x$

H $36x^3 + 14x - 10$

J $36x^3 + 42x^2 - 30x$

Name ______________________ Date ____________ Class ____________

LESSON **14-3**

Problem Solving

Adding Polynomials

Write the correct answer.

1. What is the perimeter of the quadrilateral?

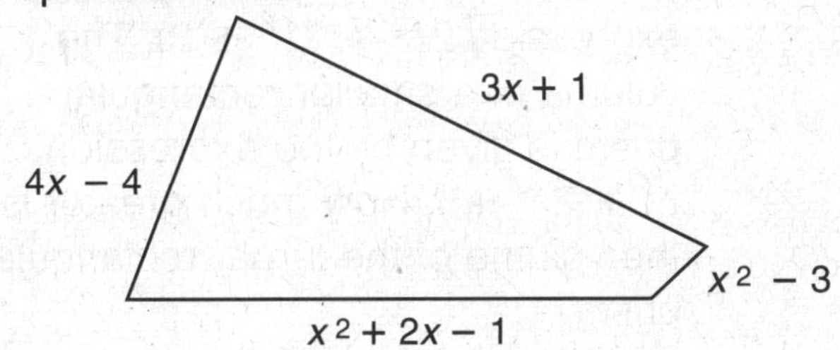

____

2. Jasmine purchased two rugs. One rug covers an area of $x^2 + 8x + 15$ and the other rug covers an area of $x^2 + 3x$. Write and simplify an expression for the combined area of the two rugs.

____

3. Anita's school photo is 12 inches long and 8 inches wide. She will surround the photo with a mat of width w. She will surround the mat with a frame that is twice the width of the mat. Find an expression for the perimeter of the framed photo.

____

4. The volume of a right cylinder is given by $\pi r^2 h$. The volume of a right cone is given by $\frac{1}{3}\pi r^2 h$. Write and simplify an expression for the total volume of a right cylinder and right cone combined, if the cylinder and cone have the same radius and height. Use 3.14 for π.

____

Choose the letter of the correct answer.

5. Each side of a square has length $4s - 2$. Which is an expression for the perimeter of the square.

   A $8s - 2$

   B $16s - 8$

   C $8s - 4$

   D $16s - 4$

6. The side lengths of a certain triangle can be expressed using the following binomials: $x + 3$, $2x + 2$, and $3x - 2$. Which is an expression for the perimeter of the triangle?

   F $2x + 5$

   G $2x - 1$

   H $3x + 5$

   J $6x + 3$

7. What polynomial can be added to $2x^2 + 3x + 1$ to get $2x^2 + 8x$?

   A $5x$

   B $5x + 1$

   C $5x^2 - 1$

   D $5x - 1$

8. Which of the following sums is NOT a binomial when simplified?

   F $(b^2 + 5b + 1) + (b^2 + 5b + 1)$

   G $(b^2 + 5b + 1) + (b^2 + 5b - 1)$

   H $(b^2 + 5b + 1) + (b^2 - 5b + 1)$

   J $(b^2 + 5b + 1) + (-b^2 + 5b + 1)$

Name ______________________ Date __________ Class __________

LESSON 14-4

Problem Solving

Subtracting Polynomials

Write the correct answer.

1. Molly made a frame for a painting. She cut a rectangle with an area of $x^2 + 3x$ square inches from a piece of wood that had an area of $2x^2 + 9x + 10$ square inches. Write an expression for the area of the remaining frame.

____

2. The volume of a rectangular prism, in cubic inches, is given by the expression $2t^3 + 7t^2 + 3t$. The volume of a smaller rectangular prism is given by the expression $t^3 + 2t^2 + t$. How much greater is the volume of the larger rectangular prism?

____

3. The area of a square piece of cardboard is $4y^2 - 16y + 16$ square feet. A piece of the cardboard with an area of $2y^2 + 2y - 12$ square feet is cut out. Write an expression to show the area of the cardboard that is left.

____

4. A container is filled with $3a^3 + 10a^2 - 8a$ gallons of water. Then $2a^3 - 3a^2 - 3a + 2$ gallons of water are poured out. How much water is left in the container?

____

Circle the letter of the correct answer.

5. The perimeter of a rectangle is $4x^2 + 2x - 2$ meters. Its length is $x^2 + x - 2$ meters. What is the width of the rectangle?

   A $3x^2 + x + 2$ meters

   B $2x^2 + 2$ meters

   C $x^2 + 1$ meters

   D $\frac{3}{2}x - \frac{1}{2}x + 1$ meters

6. On a map, points A, B, and C lie in a straight line. Point A is $x^2 + 2xy + 5y$ miles from Point B. Point C is $3x^2 - 5xy + 2y$ miles from Point A. How far is Point B from Point C?

$x^2 + 2xy + 5y$ miles

A B C

$3x^2 - 5xy + 2y$ miles

   F $-2^2 + 7 + 3y$ miles

   G $4x^2 - 3xy + 7y$ miles

   H $-4x^2 + 3xy - 7y$ miles

   J $2x^2 - 7xy - 3y$ miles

Name ______________________ Date ____________ Class ____________

LESSON **14-5**

Problem Solving

Multiplying Polynomials by Monomials

Write the correct answer.

1. A rectangle has a width of $5n^2$ inches and a length of $3n^2 + 2n + 1$ inches. Write and simplify an expression for the area of the rectangle. Then find the area of the rectangle if $n = 2$ inches.

____

____

2. The area of a parallelogram is found by multiplying the base and the height. Write and simplify an expression for the area of the parallelogram below.

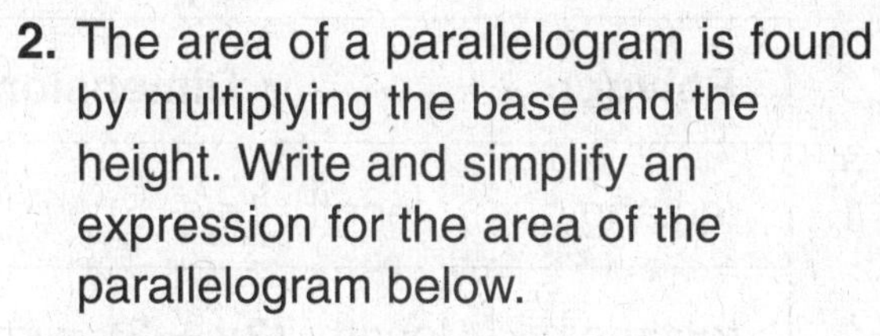

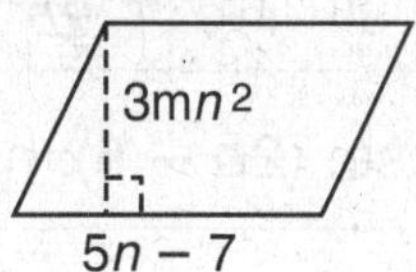

____

____

3. A parallelogram has a base of $2x^2$ inches and a height of $x^2 + 2x - 1$ inches. Write an expression for the area of the parallelogram. What is the area of the parallelogram if $x = 2$ inches?

____

____

4. A rectangle has a length of $x^2 + 2x - 1$ meters and a width of x^2 meters. Write an expression for the area of the rectangle. What is the area of the rectangle if $x = 3$ meters?

____

____

Circle the letter of the correct answer.

5. A rectangle has a width of $3x$ feet. Its length is $2x + \frac{1}{6}$ feet. Which expression shows the area of the rectangle?

A $5x + \frac{1}{6}$

B $6x^2 + \frac{1}{2}x^2$

C $6x^2 + \frac{1}{2}$

D $6x^2 + \frac{1}{2}x$

6. Which expression shows the area of the shaded region of the drawing?

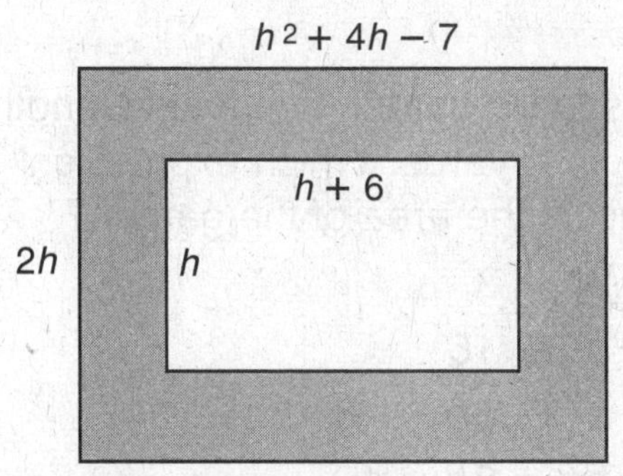

F $2h^3 + 8h - 14h$

G $2h^3 + 9h^2 - 8h$

H $2h^3 + 7h^2 - 20h$

J $2h^3 + 7h^2 - 8h$

Name ______________________ Date __________ Class __________

LESSON **14-6**

Problem Solving

Multiplying Binomials

Write and simplify an expression for the area of each polygon.

	Polygon	Dimensions	Area
1.	rectangle	length: $(n + 5)$; width: $(n - 4)$	
2.	rectangle	length: $(3y + 3)$; width: $(2y - 1)$	
3.	triangle	base: $(2b - 5)$; height: $(b^2 + 2)$	
4.	square	side length: $(m + 13)$	
5.	square	side length: $(2g - 4)$	
6.	circle	radius: $(3c + 2)$	

Choose the letter of the correct answer.

7. A photo is 8 inches by 11 inches. A frame of width x inches is placed around the photo. Which expression shows the total area of the frame and photo?

A $x^2 + 19x + 88$
B $4x^2 + 38x + 88$
C $8x + 38$
D $4x + 19$

8. Three consecutive odd integers are represented by the expressions, x, $(x + 2)$ and $(x + 4)$. Which expression gives the product of the three odd integers?

F $x^3 + 8$
G $x^3 + 6x^2 + 8x$
H $x^3 + 6x^2 + 8$
J $x^3 + 2x^2 + 8x$

9. A square garden has a side length of $(b - 4)$ yards. Which expression shows the area of the garden?

A $2b - 8$
B $b^2 + 16$
C $b^2 - 8b - 16$
D $b^2 - 8b + 16$

10. Which expression gives the product of $(3m + 4)$ and $(9m - 2)$?

F $27m^2 + 30m - 8$
G $27m^2 + 42m - 8$
H $27m^2 + 42m + 8$
J $27m^2 + 30m + 8$